STRATEGY AND PERFORMANCE OF WATER SUPPLY AND SANITATION PROVIDERS

EFFECTS OF TWO DECADES OF NEO-LIBERALISM

Strategy and performance of water supply and sanitation providers

Effects of two decades of neo-liberalism

Strategieën en prestaties van water bedrijven
Effecten van twee decennia neo-liberalisme

Proefschrift

Ter verkrijging van de graad van doctor aan de
Erasmus Universiteit Rotterdam
op gezag van de
rector magnificus

Prof. dr. S.W.J. Lamberts

En volgens besluit van het College voor Promoties

De openbare verdediging zal plaatsvinden op

vrijdag 17 april 2009 om 9 uur

door

Marco Adrianus Cornelis Schouten

Geboren te Blokker

ERASMUS UNIVERSITEIT ROTTERDAM

Promotiecommissie

Promotor:
Prof. dr. M.P. van Dijk

Overige leden
Prof. dr. R. van Tulder
Prof. dr. J.P.M. Groenewegen
Prof. dr. B. de Wit

Copromotor:
Dr. A.F. Correljé

Cover photo from Fantasy Art 3D Wallpapers at http://fantasyartdesign.com

CRC Press/Balkema is an imprint of the Taylor & Francis Group, an informa business

© 2009, Marco Schouten

Published by:
CRC Press/Balkema
PO Box 447, 2300 AK Leiden, The Netherlands
e-mail: Pub.NL@taylorandfrancis.com
www.crcpress.com – www.taylorandfrancis.co.uk – www.balkema.nl

ISBN 978-0-415-55129-8 (Taylor & Francis Group)

Abstract

This thesis concerns governmentally motivated institutional changes in the water supply and sanitation (WSS) sector, and more specifically the changes associated with the adoption of the neo-liberal agenda.

The continuous growth in the demand for WSS services has posed decision makers with the challenge to discover new, and to adapt existing, institutions. Institutional change in the WSS sector is a hazardous enterprise for any policy maker in view of the public interest at stake, the externalities associated, and the ambiguous nature of the good.

The most prominent institutional change for the WSS sector is neo-liberalism. This change that started at the beginning of the 1990s entailed essentially a call for more competition and more private sector involvement. Neo-liberalism manifests itself in the water sector through three complementary forms: a shift in ownership of the water services supplier (privatisation), enhanced competition (liberalisation), and involvement of private parties through partnership arrangements (private sector involvement).

Scholars have –to date- not succeeded in providing conclusive evidence regarding the added value of neo-liberal institutional changes to the WSS sector. The research design for the thesis at hand attempts to take an alternative approach compared to the existing body of literature, by taking 'strategy' as an intermediate variable between institutional changes and shifts in performance. The main research question of the thesis is formulated as: to what extent do neo-liberal institutional changes affect the strategies and performance of WSS providers. To operationalise the multi-dimensional nature of strategies, an analytical framework is developed distinguishing what a WSS provider can do (context), what it wants to do (plans), and what it actually does (actions).

To provide an assessment of the magnitude and rationale for neo-liberal institutional changes a case study is conducted of the Western European WSS sector investigating the drivers for change, the current institutional arrangements and plausible future developments. It is found that the current institutional context of European WSS providers is a shattered landscape of multiple, locally dependent institutional arrangements. A country-by-country analysis shows a clear shift towards more delegation and more private sector involvement. Hence, the analysis confirms that neo-liberalism has had a profound impact on the institutions of the WSS sector, and its influence will continue for the near future.

Analysing whether neo-liberal institutional changes have had an effect on what WSS providers *can* do (strategic context), results in the conclusion that institutional changes indeed make a difference. In a case study comparing the regulatory regimes of England & Wales and the Netherlands, it shows that not only the regulatory regimes are different but also that the managerial discretion differs. Managers of privately owned WSS providers enjoy more freedom with respect to their market and products/services strategies, while these companies are more constrained with respect to their seeking revenues strategies. For the internal and external organisation not much difference was found between the two comparative cases.

Analysing whether privately owned companies *want* to (strategic plans) pursue different strategies compared to public companies results in a similar conclusion. A case study situated in the Netherlands Antilles was conducted to analyse this relation between institutions and strategic plans. Again it was found that institutions matter for the strategic plans that WSS providers make. For all five strategic components the privately owned WSS provider was found to be more opportunistic, aggressive and risk taking compared to the public provider.

The conclusions with respect to the strategic context and strategic plans are partly supported by the findings from the analysis of the relation between institutions and strategic actions. This was analysed through a survey with responses from 96 senior managers over 49 WSS providers located in several Western European countries. For all strategic components statistically significant differences were found between publicly owned, mixed and privately owned WSS providers.

Hence, based on the analysis on what WSS providers with different institutional context can, want and actually do; it is concluded that institutions matter for the strategies of WSS providers. The last step of the analytical framework is to include performance in the analysis. In this regard a statistical analysis was made of the scores obtained of the strategic actions and of benchmarking indicators. For some performance indicators statistically significant relations were found, but overall they are too few and seemingly random. Hence, the inclusion of performance to institutions and conduct remains a concern. The inherent problems with performance interpretation, measurement and comparison complicate to provide any accurate insight in the value of institutional changes.

This conclusion supports the main notion of the thesis that to properly understand institutional changes, an alternative is to analyse its effect on the conduct of WSS providers. Knowing that WSS providers act differently due to neo-liberal institutional changes, gives a first inclination that also performances will be affected. In this regard the thesis opens a window for new research both on understanding better the relation between institutional changes and conduct of WSS providers, and on the relation between conduct and performances of WSS providers.

Preface

When I graduated in 1994 from the Vrije Universiteit in Economics, my supervisor Peter Tack said in his short speech that he would not be surprised if I would continue in some way in academia. At the time I didn't think much of it, yet somehow his statement remained stuck in the back of my mind. The first ten years after graduation I worked as a management consultant. I enjoyed this work tremendously as it acquainted me with a lot of different clients and analytical problems. Even I got to know different countries, as I worked and lived for Ernst & Young Management Consultants for about three years in Suriname advising a diversity of organisations like the local beer company, the oil refinery, a shrimp company, a rum producer and the Ministry of Finance. For Royal Haskoning Consulting Engineers I worked and lived in Egypt acting as the right hand of the financial director of a large water utility. In Egypt it was the first time I became involved in the water industry and I found it really enticing. When I had in 2003 the opportunity to work for UNESCO-IHE, the Institute for Water Education in Delft, I thought it would be an excellent opportunity to deepen my understanding of the sector's characteristics.

At UNESCO-IHE I found a vitalizing environment. During lunch the restaurant is like the United Nations with students from all nationalities. My prime responsibility when I joined UNESCO-IHE was to act as an administrative project leader for a large research project funded by the European Union. The project was called Euromarket, undertaken by nine European knowledge institutes over a three year period and aimed to construct scenarios for the liberalisation of the European water sector. For me, this project taught me a lot about research, partnerships, academia, the water sector and liberalisation. Next to this project, I had my regular activities as senior-lecturer in teaching and mentoring of students. Also I got involved, given my experience, in several external consultancies, like the one in St. Maarten that features in this dissertation. When joining UNESCO-IHE, I indicated that I would find it valuable to pursue a PhD, although at that time I had no clear idea about the topic or about what such endeavour entailed. However, over the years, the idea took shape and I tried as much as possible to feed regular activities into this idea.

It is said that to pursue a PhD is a lonely journey. And indeed, I found this to be true. Yet I am strongly convinced that this relative solitary state of years is not necessarily a bad thing. Even more, I believe that finding one's own path is one of the most important learning elements of the PhD process. The result of this journey you find now literally in your hands. My feelings, with respect to haven taken this journey, are best expressed by the Greek poet Kavafis in 1911[1]:

---

[1] The poem is presented in a shortened version.

When you leave for Ithaca,
may your journey be long
and full of adventures and knowledge.

Don't lose sight of Ithaca,
for that's your destination.
But take your time;
better that the journey lasts many a year
and that your boat only drops anchor on the island
when you have grown rich
with what you learned on the way.

Don't expect Ithaca to give you many riches.
Ithaca has already given you a fine voyage;
without Ithaca you would never have parted.
Ithaca gave you everything and can give you no more.

If in the end you think that Ithaca is poor,
don't think that she has cheated you.
Because you have grown wise and lived an intense life,
and that's the meaning of Ithaca.

On this journey to 'Ithaca' any PhD candidate is in dire need for guidance and support to travel successfully this path and hand-in the dissertation in time. I can gladly say that I was fortunate enough to receive such guidance and support and therefore like to acknowledge these persons in this section.

First of all, I would like to express my appreciation and gratitude to my promotor Professor Meine Pieter van Dijk. I'll never forget that the both of us went to Nelspruit, South Africa, to train local politicians on the topic of public private partnerships. Every morning, near the side of the pool, we shoved the remainders of the breakfast aside, discussing another of the draft chapters. This week in South Africa surely made a breakthrough. Next, I sincerely thank my co-promotor Aad Correljé. I met Aad during the Euromarket project, and I really profited a lot from his kind advice and stimulating conversations. I am also thankful for the critical and helpful comments that I received from the members of the promotion commission.

Next, I want to express my gratitude to my employer UNESCO-IHE, the Institute for Water Education. UNESCO-IHE gave me the freedom and opportunity to pursue the PhD. In this respect I especially like to mention Frank Jaspers, who allowed me to minimize my foreign travels in the last year of writing, which was crucial to keep up the pace. With Klaas Schwartz, one of the members of the Water Services Management Core, I authored several articles; and with Damir Brdjanovic I worked closely together on the St. Maarten assignment, resulting also in an article. Also I had the pleasure to mentor several students that contributed in the execution of the research. I particular I like to thank Gaetano Casale, Maria Pascual Sanz and Hilman Agung for their work.

Since the research is situated in the water sector, I am indebted to many experts working in the sector. Especially I am grateful for the support I received from Vitens, through Jan Hoffer and his colleagues, from Suez Environnement through Jacques Labre and his colleagues, and several key informants in the Dutch water sector.

Then, I would like to thank my family and friends for being supportive all along. I particularly thank my parents for their love and support. Being a parent now myself, you have set an example worth imitating. I can only hope that my daughters will find me as good a parent as I think the both of you are. Also I thank my three beautiful daughters Louise, Sofie and Josette for the joy they give me. And I'd like to single out my two friends, Edwin and John, for agreeing to act as paranimf during my defence. I am happy you want to stand by me at this special occasion.

Lastly, I thank Nicole. Nicole, for some odd reason you believed in me along the way, while at the same time reminding me there is so much more to life than this. I love and thank you for that (amongst other things).

Table of Contents

Annexes

List of Figures

List of Boxes

List of Tables

List of Graphs

List of Scatter Box Diagrams

Section I: Introduction

Section I. Introductory section

Chapter 1 Introducing the Thesis

This first Chapter provides an introduction to the thesis by presenting the topic and structure of the thesis.

1.1   Introduction

The water supply and sanitation (WSS) sector finds itself in the limelight nowadays. The increased visibility of the WSS sector has several reasons, which all can be largely traced to the increasing demand for the product that the sector generates: drinking water. The need for water is felt more harshly as populations continue to increase, putting the supply of sufficient quantity and quality of water in the centre of attention. Since the 1950s world population has doubled and water use has even tripled; yet the quantity of available fresh water remains equal to the amount one million years ago (Dalhuisen *et al.*, 1999). The demand for water is expected to continue to increase as the world's population will further grow from 6.5 billion today to 9.1 billion in 2050 (UN, 2007).

Over the last two decades the international WSS sector has witnessed major institutional changes. The large-scale adoption of the neo-liberal agenda by national, regional and local policy makers dramatically changed the institutional landscape of the WSS sector. The increased involvement of private parties and the stimulation of competition implicated a pronounced shift in the traditionally public and monopolistic character. This shift has spurred a body of research on the value and effects of the neo-liberal institutional changes. To-date, despite the large quantity of studies, the available empirical evidence is less robust than one would hope for, both in quality and in scope.

This thesis can be considered as a contribution to the existing body of knowledge on the effects of neo-liberal institutional changes in the WSS sector. It particularly contributes by using an alternative approach compared to the majority of existing studies. The research objective is achieved by undertaking a research approach based on notions from New Institutional Economics (NIE) and strategic management literature. Through the use of a testable analytical path model the relations between institutional changes, changes in conduct (strategies), and changes in performance of WSS providers are better understood.

1.2   Structure of the thesis

The structure of the thesis is subdivided in 4 sections. In addition to the executive summary and the foreword, the following sections are incorporated: (I) introduction, (II) research design, (III) analysis, and (IV) conclusions. In total the 4 sections comprise 11 chapters.

In most of the chapters purposely a case study is included. According to Yin (2003) case studies may be instrumental for several objectives. In the chapters of the Introductory Section the prime objective for inclusion of the case studies is illustrative. The Introductory Section intends to show the relevancy of the research both from a societal as from a scientific point of view. Therefore the case studies in these chapters serve to show how in real life policy makers, decision makers and scholars alike have struggled with the focal problem of the thesis; e.g. the lack of insight in the effects of neo-liberal institutional changes in the WSS sector. In the chapters of the Analysis section the case studies serve a different objective. In these five chapters the inclusion of the case studies is for analytical purposes. These case studies are intended to test one or more of the hypotheses of the research design.

Figure 1 below depicts the structure of the thesis:

Figure 1 Structure of the thesis

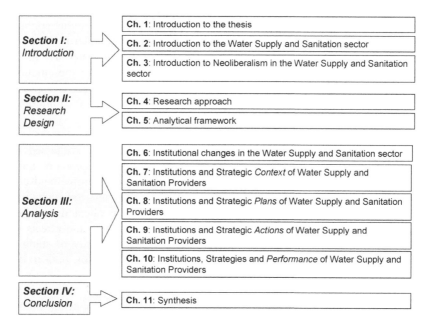

I Introductory section

This first section is composed of three Chapters. Chapter 1 is meant to introduce the theme of the thesis and to provide a structure to the thesis. Chapter 2 introduces the WSS sector. Two main findings are identified: first, institutions have a large role to play in the WSS sector due to the nature of the good, and second, traditionally the nature of the good has led to a dominant role for governments both as responsible and management entity. An illustrative case study from Friesland, the Netherlands, is added to emphasize the traditionally public character of the service provision. Chapter 3 introduces neo-liberalism in the WSS sector. The theories and rationale underpinning the implementation of neo-liberalism are presented. An illustrative case study from Cochabamba, Bolivia serves to sharply highlight the lack of understanding of decision makers regarding the implications of adopting the neo-liberal agenda.

II Research design section

The section on research design is composed of two Chapters. Chapter 4 builds on the Introductory Section by outlining the research aims and approach. Specific attention is given to explaining the construct of strategies, as it features dominantly in the thesis. Chapter 5 completes the Research Design section by laying out the consequences of the choice to select strategies. Research questions, hypotheses and techniques are described here. Particular attention is provided on the limitations of the research.

III The analysis section

This section is composed of five Chapters each analysing at a different level the neo-liberal institutional changes in the WSS sector.

Chapter 6 assesses the extent to which neo-liberalism has impacted, and will impact, the institutions of the WSS sector. The drivers for implementing neo-liberal institutional changes are assessed, as well as the current and future institutions for the water sector. Such is conducted through a case study of the Western European WSS sector. The Chapter concerns itself with identifying both the magnitude as well as the drivers of institutional changes in the WSS sector.

Chapter 7, 8 and 9 each analyse one of three dimensions of strategy, e.g. respectively the strategic context, the strategic plans and the strategic actions. Chapter 7 analyses the strategic context of WSS providers. The guiding question for this Chapter is: is there a difference in what WSS providers in different institutional contexts *can* strategize, or are they in a straightjacket of regulatory and sector impositions. In this order a comparative case study analysis is made of the (neo-liberal) regulatory regime in England and Wales, and the (traditionally public) regulatory regime in Netherlands. Chapter 8 provides the second level of analysis as a comparison is made whether a publicly owned provider has different strategic plans compared to a privately owned provider. The main guiding question for this Chapter is: is there a difference in what WSS providers in different institutional contexts *want* to pursue as strategies. In this order, a single case study analysis is conducted of the strategic plans of both a publicly owned and of a privately owned operator, as made explicit in the bidding documents for a water concession on the Island of St. Maarten of the Netherlands Antilles. Chapter 8 investigates whether providers with different ownership characteristics have different strategic actions. The main guiding question for this Chapter is whether WSS providers in different institutional contexts *do* have different strategies. This is researched through a survey among water utilities in several Western European countries.

The final Chapter 10 of the 'Analysis'-section aims to build on the results of the previous analytical Chapters by establishing a relation between neo-liberal sector reform measures, strategies of operators, and performance of service operators. In this order an analysis is made between strategic actions (as identified in the previous Chapter) and performance.

IV Conclusion section

The conclusion section is composed of one Chapter combining the discussion and conclusion related to the findings presented in the thesis. In the conclusion a reflection is made whether the research aims and objectives are met. The contribution to knowledge through the study at hand is shown and areas for future research are identified.

Section I   Introductory Section

Chapter 2 Introducing the Water Supply and Sanitation sector

This Chapter provides an introduction to the WSS sector by identifying the main characteristics of the WSS sector and the role institutions play in the sector.[2]

2.1   Introduction

This Chapter introduces the WSS sector. WSS services are defined and distinguished from other water services, like hydropower and irrigation. The main characteristics of WSS services are identified, followed by an overview of the main institutional arrangements in the sector. One of the main findings from the Chapter is that many of the institutions and of the contemporary developments in the WSS sector may be (partly) explained by the nature of WSS services.

2.2   Characterizing WSS services

Water serves many uses, like for drinking and hygiene, but also to grow crops, to generate electricity, to navigate boats, and for recreational purposes. Hence, WSS services are only one part of the more general term 'water services'. The European Water Framework Directive (WFD) defines water services in the following manner (WFD, 2000: Article 2, point 38):

> Water services are all services that provide, for households, public institutions or any economic activity: (a) abstraction, impoundment, storage, treatment and distribution of surface water or groundwater, (b) wastewater collection and treatment facilities which subsequently discharge into surface water.

Examining this definition, water services incorporate both the activity to use water for irrigation, water transfer, water for hydropower, and drinking water supply and sanitation. Moreover, according to the definition it does not matter whether individuals or third parties provide the service. Hence also self-supply by individuals is included. The term 'water supply and sanitation services (WSS)' is more restrictive as it only concerns the supply of drinking water and the collection and treatment of wastewater by authorized WSS providers. Self-supply is excluded. WSS services relate to the 'small' cycle. Water is abstracted by an authorized provider from a river, an aquifer or in some cases even the sea. This water is treated and pumped into underground pipes, ending up at the premises of consumers where it flows out of their taps. The wastewater that comes from the shower pit, the latrine and sometimes from the drains in the street, flows into another underground piping system, ending up at a

---

[2] Parts of of this Chapter (including the Friesland case) have been published in Schwartz, K. and M. Schouten, 2007. *Water as a political good: revisiting the relationship between politics and service provision.* In: Water Policy. Volume 9. Number 2, pp. 119-129. IWA Publishing. Also some parts have been published in Fonseca, C., Schouten, M.A.C. and R. Franceys, 2005. *Plugging the Leak. Can Europeans find new sources of funding to fill the MDG water and sanitation gap?* Published by IRC. June 2005. 54 pages.

wastewater treatment plant. There, the materials that really harm the environment are removed before the wastewater is discharged into the environment. WSS providers, all over the world, have managed this cycle for more than 100 years, and the fundamentals of the processes remain largely untouched (Thomas and Ford, 2005).

Several authors suggest that the WSS sector has characteristics that make it relatively unique compared to other sectors. Pargal (2003: 23) based on an econometric assessment of private investment flows and data from Latin America concludes that:

> the water sector differs materially from [telecom, electricity and road]...: private investment in water is not significantly affected by the passage of reform legislation in the sector and public expenditure is very important and only mildly substitutable for private spending.

The unique character of the WSS sector is often argued by pointing out the diverse characteristics of water. For example Savenije (2002) identified a long list of characteristics underlining the special position of water, being: water is essential, water is scarce, water is fugitive, water is a system, water is bulky, water is non-substitutable, water is not freely tradable and water is complex. The combination of characteristics makes the provision of WSS services as a class of its own, and also with problems and solutions of its own. Individually these characteristics are applicable to many goods and sectors, but their combined applicability makes the WSS sector unique from other sectors[3].

However the uniqueness of the WSS sector is also questioned. For example, Briscoe (1997) rejects the notion of the uniqueness of the WSS sector compared to other sectors and addresses this as the 'exceptionalism syndrome'. Briscoe observes that in many cases the supposedly unique character is used as a universal argument against change in the WSS sector. He, conversely, advocates that there is a remarkable degree of commonality between the ingredients of successful reforms in quite different contexts. The institutional arrangements may be different in different locations, but the underlying principles are similar, in Briscoe's perception.

Despite the critique of Briscoe, the following sections will address the extent to which WSS services are indeed unique. It will elaborate on the ambiguities related to the nature of WSS services, by addressing a series of ascribed characteristics.

2.2.1   A public and/or a private good

The first ambiguity with respect to WSS services, which both policy makers and practitioners need to cope with, is whether to perceive it as a public or as a private good. This perception is important as it may have consequences for the type of institutions government may favour.

Traditionally, WSS services provision is conceived as a 'public good' (Aalbers *et al.*, 2002). However the term of "public good" has different meanings and can therefore lead to confusion. In some cases "public good" refers to the fact that a service is

---

[3] Memon and Butler (2003a) tried to consolidate the different characteristics of water by identifying it as a 'social good'. They interpreted water to be one of the basic ingredients needed to sustain life, and therefore has historically been viewed as a social good. In their view, reasonable governments remain under moral obligation to ensure safe and affordable provision of water related services. According to Memon and Butler (2003a) the special nature of water is enhanced by its religious dimension (e.g. baptism in Christianity, wadhoo in Islam). This strong emotional affiliation, affordability issues, plus social and political drivers have largely caused the interventions of the public authorities in the provision of WSS services.

offered to the general public; in other cases the term highlights that a service has been assigned a specific role in the public interest; or "public good" can refer to the public ownership of the entity that is providing the service. The neoclassical definition of what constitutes a public good is far more restrictive (Van Dijk, 2003). The essential properties of public goods are according to neoclassical economic theory:

1. Non-excludability; that if the good is provided for one person, it is automatically available for everybody else. In practice excludability is determined by an assessment of economic costs and benefits. Only in exceptional circumstances it is technically not possible to exclude individual consumers from drinking water and sanitation. However, the exclusion of consumers from water is restricted legally (i.e. there is an obligation to supply and a ban on disconnections). This is because some governments have determined that the economic costs (regarding public health, social obligations) outweigh the economic benefits (regarding preventing free riding) of exclusion.

2. Non-rivalry; meaning that the good is not less available for any one person because another person is enjoying it. The extent of rivalry of WSS services is more difficult to determine, as it will partly depend on infrastructure capacity considerations. These considerations may change over time – both in the short run as a result of regular fluctuations in demand and in the long run as the physical capacity of the existing infrastructure is reached. There is an argument that WSS provision exhibits both high and low rivalry during any given year.

Hence, according to neoclassical economic theory, excludability and rivalry should dictate whether collective provision is made for free to the individual consumer with the cost financed out of general taxation. It is evident that in a strict use of economic theory WSS provision cannot be considered as public goods or common pool goods. WSS provision is usually characterised by excludability and, at certain times of the year, rivalry. For WSS services it is feasible and in fact common practice both to charge users and to exclude non-payers. Although collectively provided, there are few cases in the world where WSS services are provided free-of-charge as is normally the practice for pure public goods. Hence, in the interpretation of neoclassical economic theory WSS services are private goods (Nickson, 1997).

2.2.2   A monopoly and/or a competitive good

Another ambiguity related to the ascribed characteristics of WSS services is that often the whole sector is defined as monopolistic. However, taking apart the process for WSS provision it is possible to identify some elements that are competitive, while other elements can be found to be monopolistic.

Natural monopolies are any economist's nightmare as the simplest definition of economics is "the application of reason to choice" (Green, 2000). A natural monopoly is defined as: a situation where for technical or social reasons there cannot be more than one efficient provider of a good[4]. What really differentiates a natural monopoly from a competitive market is "capture", being the absence of competitive alternatives. In a natural monopoly there is no possibility to choose. A natural monopoly does *not*

---

[4] Characteristics of a monopolistic market are: (i) The monopolist is the only supplier in the market with many consumers; (ii) At the market a homogenous product is supplied; (iii) The market is transparent. There are no competitors and the consumer has full information on the monopolist and its product; (iv) The market is closed since the supplier has a monopoly; he is the only one that can market the product (Green, 2000).

arise from government intervention in the marketplace to protect a favoured firm from competition but rather from special characteristics of the production process in the industry under the current state of technology. In this respect the WSS sector seems to fit this profile since it is extremely capital intensive (Dalhuisen *et al.*, 1999). Huge investments have been made and will remain to be made in developing and maintaining the network used to transport water, and in building facilities to collect and treat water from the various sources. The WSS sector is characterized by a high degree of sunk costs as fixed costs of water infrastructure generally make up two-thirds to 80 percent of the costs of supplying services (Noll *et al.*, 2000; Kessides, 2003). In addition, the WSS sector is subject to large economies of density, meaning that for a given distribution network, increasing the number of households connected or their consumption reduces the network's average costs (Spiller and Savedoff, 1999). This means that the provision of WSS services would be subject to declining marginal cost for service provision. These two factors cause distribution of water supply to be "a natural monopoly bottleneck to an urban water system" (Noll *et al.*, 2000).

Views about natural monopolies have altered radically in recent years. Before it was often assumed that each sector in its entirety constituted a natural monopoly. All WSS services can indeed be provided by one organisation but this is not necessarily so, and even quite uncommon[5]. WSS services are not an amorphous whole but can be divided into numerous autonomously managed activities: as water treatment - local water distribution – sewage collection – sewage treatment. The definition and allocation of exclusive rights between these activities, as well as the cross-subsidization, have become key issues. Several of these activities are potentially competitive: only the distribution networks of pipes are genuine natural monopolies.

2.2.3   An economic and/or a merit good

The third ambiguity refers to the claims to identify it as a merit or as an economic good. However, looking more in depth at the definitions of merit and economic goods, it is found that they are not contradictory but more complementary.

Teulings *et al.* (2003) specifically state that the provision of WSS services is a merit good, arguing that the public interest is at stake in the provision of the service. Two conditions are to be met according to Teulings *et al.* (2003) for the public interest to be at stake: (i) externalities, and (ii) free-rider behaviour. An externality is when due to the action of one party or a transaction between two (or more) parties, a third party is facing positive or negative effects. The more there is demand for drinking water, the more the strain on the natural environment through increased abstraction rates, and larger wastewater discharge quantities (Glaister, 1996; Jeffery, 1994). Economic externalities arise as WSS providers are employers and charge the population for consuming their water. Especially in poor areas attention needs to be dedicated to ensure that drinking water remains affordable to prospective consumers, putting issues as ability-to-pay and equity in the spotlight. Even the maintenance of the network of the WSS provider might bring severe social and economical externalities, such as traffic disruptions, bursts and collapses. Free rider behaviour is when some of the involved parties refuse to contribute to the action/transaction, because they are

---

[5] For example, the three separate water management entities of the city of Amsterdam became as of 2006 the first integrated water service organisation in the Netherlands, claiming 5% cost savings due to the prospected merger (VNG, UvW and VEWIN, 2006).

speculating that the result of the action/transaction will also be realised without their contribution.

The International Conference on Water and the Environment: Development Issues for the 21st Century held in Dublin in 1992 provided additional impetus by recognizing water as an 'economic good'. One of the Four Dublin Principles was stated as:

> Water has an economic value in all its competing uses and should be recognized as an economic good. ... Managing water as an economic good is an important way of achieving efficient and equitable use, and of encouraging conservation and protection of water resources.

At that time this recognition was fairly controversial. The reason for the controversy is due to possible conflict with the merit good characteristic of WSS services. The identification of water as an economic good, highlights that water is scarce and has an economic value in all its competing uses. Although everybody would agree that water is indeed scarce and has competing uses, the label of being an "economic" good has arguably been the most contested and debated of the Dublin principles. It sharply focussed discussion of WSS provision to questions relating to the financing of the service provision process and to new economic instruments that could be explored to realize service provision improvements.

2.3   Institutions in the WSS sector

The previous section makes clear that many, sometimes conflicting, characteristics are ascribed to WSS services. Governmental policy makers determine the institutions for WSS providers based on their interpretations of the nature of WSS services.

Phrased in the terminology of New Institutional Economics (NIE) the main task for policy makers is to establish the most optimal institutional context to adequately respond to the growing demand for WSS services. According to Williamson (1998), NIE is a body of thinking based in two propositions: institutions matter and institutions are susceptible to analysis. Institutions are in the interpretation of NIE "the rules of the game in a society, or more formally, are the humanly devised constraints that shape human interaction" (North, 1990: 3)[6]. Policy makers have a primary responsibility in shaping the society's institutions in an effort to secure the public interest. Institutions, and the way they evolve, shape human behaviour, and subsequently affect performance. When the rules of the game change, the players play differently and the game may have a different outcome.

Institutions are composed of (in)formal rules or constraints, and of their enforcement characteristics. Formal rules include political (and judicial) rules, economic rules, and contracts. Political rules broadly define the hierarchy of the polity, its basic decision structure, and the explicit characteristics of agenda control. Economic rules define property rights. Contracts contain the provisions specific to a particular agreement in exchange. Informal constraints are extensions, elaborations and qualifications of rules that 'solve' innumerable exchange problems not completely covered by formal rules and that in consequence have tenacious survival ability. They allow people to go about the everyday process of making exchanges without the necessity of thinking out

---

[6] Institutions differ from organisations. Institutions are the rules of the game; organisations are groups of individuals bound together by a common objective function.

exactly at each point and in each instance the terms of exchange. Routines, customs, traditions, and culture are words often used to denote the persistence of informal constraints. They include conventions that evolve as solutions to coordination problems and that all parties are interested in maintaining (such as traffic rules for example), norms of behaviour that are recognized standards of conduct (such as codes of conduct that define interpersonal relationships in the family, business, school, etcetera), and self imposed codes of conduct (such as standards of honesty or integrity). Conventions are self-enforcing. Norms of behaviour are enforced by the second party (retaliation) or by a third party (societal sanctions or coercive authority), and their effectiveness will depend on the effectiveness of the enforcement (North, 1997).

Matthews (1986) presumes that in the course of time people have discovered and adopted institutions that enabled them to co-operate with one another more efficiently than they did before. On this reckoning, institutional changes are a necessity for economic growth. While the polity can change formal rules overnight, informal constraints change very slowly. According to North (1997) there is frequently a significant difference between intended outcomes of an institutional change, and actual outcomes. Outcomes frequently diverge from intentions because of the limited capabilities of the individuals and the complexity of the problems to be solved. In his concluding remarks, Matthews (1986: 917) brings out clearly the difficulties to understand the implications of institutional changes:

> Politicians of all parties are great believers in institutional change as a source of economic improvement – not surprisingly, because that is the sort of change they are well placed to bring about. Economists give them advice to this end, often with no less enthusiasm. Yet among the main features of institutional change is its complexity and the unforeseeable nature of its consequences, setting us off on random walks to goodness knows what destination. ... if we are to abide by scholarly standards, we have to recognize candidly that institutional changes can easily lead in the long run to results that are quite different from intended – rather in the same way as wars have been found to do.

The WSS sector is known to be one of most regulated worldwide (Robinson, 1997); hence, (formal) institutions play a major role in this sector. The high degree of regulation is an indication of the importance for policy makers to have a good insight in the relevancy of the institutions they develop. According to the Dutch Advisory Council for Governmental Policy (Wetenschappelijke Raad voor het Regeringsbeleid, 2000) policy makers have four instruments at their disposal to shape the institutional context of WSS providers. The first instrument of which one might think of is by drafting laws or contracts that define the legal framework for a WSS provider to operate in. A less formal instrument compared to legislation is if the government attempts to change the 'institutional values' of WSS providers. This entails that the government may purposely create a climate in which WSS providers are automatically geared to serve the best interest of the government. A third instrument available to policy makers, is by getting involved themselves in the service provision, e.g. by making use of a hierarchical relation. Traditionally, the use of this third instrument has been popular. These instructions are not laid down in laws or contracts but are very case-specific. The fourth instrument available to policy makers is by making use of the market mechanism. Government in this respect can implement types of competition in the market, or for the market, aiming to secure the public interest associated with these gains. Each government needs to select its' own mix of this set of instruments, dependent on the local situation and preferences. 'One-size-fits-all' solutions for WSS sector institutional changes are seldom available as local circumstances differ from region to region (World Bank, 2006).

In an empirical enquiry on the variety of institutional arrangements in Western European countries, Eureau identified a number of commonalities in institutional arrangements (Eureau, 1993 and 1997). Eureau's focus on a limited number of commonalities reduced the large diversity of institutional arrangements, although it needs to be acknowledged that arrangements in the same class might show significant differences. From the classification of Eureau two main commonalities are filtered that shape the institutional arrangement:

1. Direct or delegated management. The character of water and sanitation implies that the government assumes the ultimate responsibility for service provision but how it arranges the management of the service provision is up to the government to decide upon. For one, it can choose to execute the management of the service provision itself, with a very limited degree of separation between the government and the service provider. Another option is that it relies on an Agent to execute at arms' length the management of the services, providing the Agent a certain level of autonomy.

2. Public or private management. Another dimension that reflects the institutional arrangement is if the service provision is managed by public or by private parties. The government might choose to involve a private sector to take upon themselves the management of service provision bringing with it commercial attitudes, external financial resources and know how.

Four main generic institutional contexts emerge from the classification:

1. **Direct public management**, the government is responsible for service provision but also chooses to execute the management tasks. In this case there exists no contract between the government and the services provider. Even more, it is hard to make a distinction between the government and the services provider.
    a. Organisational autonomy: Most of the time this is shaped in a format where the management of the WSS services is undertaken by a department of the municipality. The executing entity has very limited autonomy.
    b. Tariff setting: mostly tariff setting is conducted by the government.
    c. Access to funds: The service provider normally has no access to external, and needs to rely on the government for funds for investments or even operations and maintenance.
    d. Ownership: the government normally owns the infrastructure.
    e. Regulation and control: since the service provider has very limited autonomy the regulation and control is very direct from the government to the WSS service provider.

2. **Delegated public management**, the government appoints a service provider to execute the management of WSS services at arms' length. Often several neighbouring municipalities combine the execution of WSS services for a region in one service provider.
    a. Organisation: the service provider takes the form of a separate public company operating at arms-length of the government. Although the service

provider is (partly) in the hands of the government, it is able to operate relatively autonomous.

    b. Tariff setting: mostly tariff setting is delegated to the service provider.

    c. Access to funds: service providers as semi-autonomous entities have the ability to access funds outside the government, although subsidies are common for major investments.

    d. Ownership: the government normally owns the infrastructure.

    e. Regulation and control: the government is acting as a shareholder in the company, to be able to control the management of the tasks. In some cases the government allows minority shareholding of private parties.

3. **Delegated private management**, the government appoints the management of tasks to be conducted by a private entity on the basis of a temporary contract, for example a concession contract. The private entity in this respect assumes the role of the service provider.

    a. Organisation: the private entity is independent from the government and acts as a contractor.

    b. Tariff setting: mostly this is arranged in the contract.

    c. Access to funds: funding is arranged depending on the type of contract. If the contract is a lease contract, the government will need to provide the investments. If the contract is a concession, the funding is delegated to the private party. Obviously the private party can access funds as it chooses.

    d. Ownership: the government owns the infrastructure.

    e. Regulation and control: most of it is arranged in the contract, otherwise mediators and courts will be asked to solve conflicts.

4. **Direct Private Management**, in this case the public authorities limit themselves only to control and regulation. All tasks, responsibilities and ownership are placed in the hands of private parties. In this case the private party becomes the outright owner. This is the most radical change as it often runs up against the immediate limits of practicality and public acceptability. There is not really a contract between the government and the service provider, but more of a license to operate.

    a. Organisation: the private party assumes full responsibility and is independent from the government, apart from regulatory issues.

    b. Tariff setting: the service provider does tariff setting although mostly controlled and regulated by the government.

    c. Access to funds: the private service provider can access funds as it chooses.

    d. Ownership: the private service provider owns the infrastructure.

    e. Regulation and control: an extensive system of regulation and control will be set up to guarantee the public interest.

These four institutional arrangements reflect the degree of separation between the government and the service provider, which generally decreases in strength from 1 to 4, or put another way the extent to which the government has handed control of funding investment and management decisions to the service provider.

Figure 2 Governance and regulatory interface

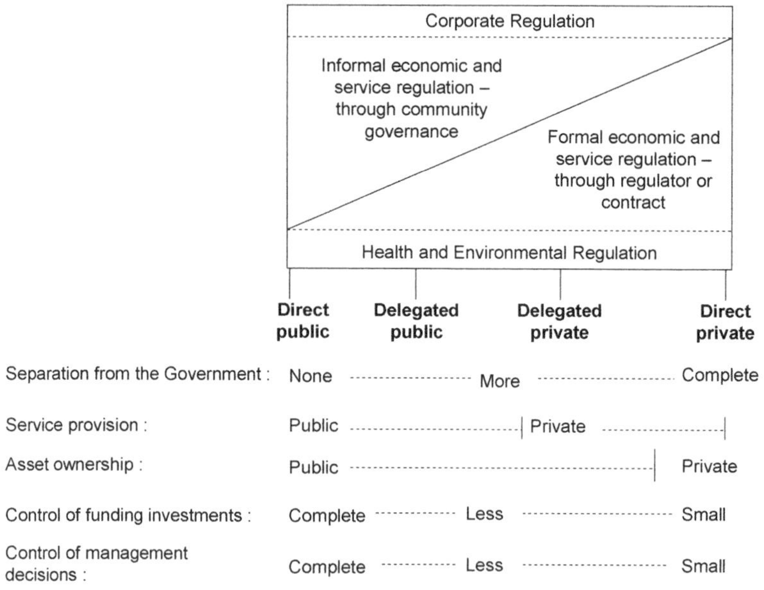

Source: IWA (2003); modified by author.

For all four institutional arrangements, a system of regulation –informal or formal- is required to ensure that the public interest is served properly (Aalbers *et al.,* 2002). Regulation prescribes WSS providers on the type and quality of the products or services to be produced, the people they want to sell it to and the price they can levy for their services. Even the amounts of water they can abstract, or any other relevant issue that the regulatory body chooses to be relevant, can be tightly regulated. These regulatory measures consequently severely constrain the latitude of water providers to engage into strategic and operational actions of their choice (Robinson, 1997; Carney, 1990). Figure 2 underlines that the regulation becomes more formalized as the WSS provider is distancing itself from the government. In direct public management, regulation is accomplished largely through community governance (political oversight) rather than explicit regulation, while in delegated private management, the regulation is largely embedded in the terms and conditions of the service contract, which is negotiated with and endorsed by the government. For the most distanced arrangement (direct private management), regulation is typically accomplished through formal and independent regulatory authorities.

2.4   The benevolent role of the government in Friesland[7]

The case of service expansion in Friesland highlights, first and foremost, the crucial role of the municipal governments. The municipalities mitigated the revenue risk of the water company in which they themselves became shareholder. With these guarantees for a minimum income, the water utility was able to secure loans for expanding its services. The municipalities also passed regulations, forcing households

---

[7] This case description is a revised version of a case study by Schwartz and Schouten (2007).

to connect to the service network and actively sought to provide incentives for the citizens to switch to the piped network. What is noteworthy in this respect is that the provision of guarantees, and even the subsidies for expanding service coverage to 'non-profitable' areas was done without introducing price distortions, in the sense that the tariffs paid by the consumers were never subsidized. Rather subsidies for service expansion went directly to the utility, which charged cost-recovering tariffs to the consumers since its establishment in 1922.

In the Netherlands, the development of safe drinking water supply was left to local initiative. Although the 1867 'Report to the King' described the generally poor condition of water supply throughout the country and emphasized the need for a national initiative, this advice was not heeded. It would be another 40 years before the Dutch government would actively involve itself with water supply. Until that time local governments and private entrepreneurs established piped water supply systems. They did so particularly in the larger and richer municipalities, where attractive rates of return on investment could be achieved. The necessary capital was made available either from municipal budgets or provided by local, British and Belgian financiers. This meant that the development of WSS services showed a strong urban bias, with the provision of the rural municipalities staying behind.

From 1910 onward this started to change. For the first time, funds were allocated to water supply at the national level, and in 1913, a permanent advisory committee to the government and a national bureau were established to advice on and assist with drinking water supply development. Their concern was mainly with the development of rural and particularly regional systems. These national initiatives, together with the reinstatement of financial autonomy for the provinces in 1905, ending what is known as the 'century of the municipality', created the necessary administrative and professional capacity at the supra-municipal level to further the extension of water supply coverage to the rural areas - a development that would take more than 50 years to be completed.

The development of water supply services in Friesland started 35 years after the first Dutch water supply company (in Amsterdam) became operational. The start-up of water supply services in 1889 had been preceded by heated debates in the municipal council of Leeuwarden town, the capital of Friesland. Some council members objected to taking an initiative in public water supply on the grounds that it would create a precedent. If they conceded to get involved in providing WSS services, how would they be able to stem the flood of expected requests for other public services and issues? In 1884, however, a decision was reached to award a municipal concession for public water supply. The concession, for a period of 50 years, stipulated service level, quality and price of water, and financial arrangements between a private company, called the Leeuwarder Waterleiding Maatschappij (LWM), and the municipality of Leeuwarden. Under this arrangement the municipality was to subsidize operations for 14 years with a gradually decreasing subsidy against which the company supplied up to 25,000 cubic meters per year free of charge to municipal buildings and public stand posts for the needy. After this period, the municipality would have to pay for its water, but would also receive a 20% share of the profits.

The LWM did quite well in its early years. After 1915, during World War I and immediately thereafter, company profits started to decrease. Rising prices of coal and chemicals drove up operating costs. The company tried to economize on expenditures by reducing the dosing of chemicals and by encouraging the municipality to proclaim a street-scrubbing ban. A proposal on the part of the company to raise tariffs was refused by the municipality on the grounds that the company was not meeting the conditions of supply specified in the concession. After 1917, dividends dropped to zero, so that the municipality also stopped receiving a profit share. It also became clear that the owners would not be willing to invest in new infrastructure, which was forecasted to yield no dividend for at least 10 years.

In 1921, the concession for the supply of Leeuwarden was withdrawn from LWM and water supply came under municipal control. The municipal councillors knew that considerable investments were required to remedy persistent supply problems and planned to share this financial burden with other municipalities.

In 1922 a regional water supply company, the 'Intercommunale Waterleidingmaatschappij Gebied Leeuwarden' (IWGL) was established by the town of Leeuwarden and eight surrounding municipalities. The utility was established as a government-owned company, which meant that it operated under company law, similar to a private company, whilst the shares of the company were in hands of local government authorities. At the time of its foundation the utility had the specific mission to not only serve the towns and villages located in its service area, but also the surrounding rural areas.

The expansion of the water supply system was financed primarily by loans that were extended on the grounds of projected company income from the new supply areas. This income was secured by means of contracts for the supply and sale of water that were drawn up between the company and newly participating municipalities, who also became shareholders in the water supply company. Similar contracts were drawn up with prospective rural industries with large water needs, particularly dairy factories. In these contracts the municipalities and the factories guaranteed a minimum use of water against a set price. In essence, what the municipalities were doing was to reduce the revenue risk for the water supply company by guaranteeing a minimum income, allowing it to obtain loans for service expansion.

Whereas the factories incorporated the water costs in the price of their product, the municipal councils had little alternative but to pass on their bills to the citizens of their municipality. This meant that the municipality had to try to encourage the citizens to connect to the network and to consume sufficient water in order to cover the off-take agreement between the municipality and the water company.

The first challenge (getting people to connect) was mainly attempted by passing municipal regulations requiring house owners to connect to the piped water supply system. Although this substantially increased the number of connections, it did not necessarily achieve the second challenge (sufficient consumption to meet the off-take as guaranteed in the contracts). Even with the connection in place, many users preferred traditional sources, such as open wells and drainage ditches, which these users had been using for many years.

In order to get consumers, who were connected, to actually start using the water supplied through the network the municipalities undertook two main activities:

- Campaigns were organized, encouraging the use of piped water. These campaigns included village visits by the water supply company director, who, in those days, was a dignitary and sure to bring out a large portion of the village population.
- On the commercial side, the tariff system was altered with the aim of raising revenue and inducing consumption. The water bill essentially consisted of two parts: a fixed part that every connected household had to pay and a variable part, which depended on the volume of water consumed. The change in the tariff system consisted of raising the fixed part of the bill and lowering the price per cubic meter of water consumed. With connected households having to pay the fixed part of the bill regardless of their consumption, the new tariff system provided an incentive for consumers to switch to the piped water network.

In 1931, the number of participating municipalities had risen to 16 - out of a total of 44. The combined area of these municipalities encompassed approximately 40% of the province's surface area. By 1947, about 50% of the provincial population was covered. In 1962, coverage had reached 95%. By the end of the 1960s coverage was 100%. The rapid expansion of coverage between 1945 and 1970 was facilitated by national subsidy schemes for capital investment enabling the construction of otherwise non-viable extensions to the most remote rural villages and hamlets. 'Non-profitable' expansion of the network was eligible for subsidy by the national government (70%), the provincial government (10%), and the municipal government (10%). Subsidies were paid into a sinking fund to pay for future expansion.

In this period, in addition to expanding coverage in the existing supply area, the IWGL also absorbed the two remaining independent urban water supply systems. In 1954 the water supply system of Heerenveen was taken over, and from 1959 onwards, IWGL supplied water in bulk to Sneek. The full take-over of the Sneek system followed in 1977. Pre-empting this final action, IWGL was renamed 'N.V. Waterleiding Friesland' (WLF) in 1974, signifying that its supply area from then on was the entire province of Friesland. The achievement of universal coverage in the late sixties signalled the end of a period of unprecedented growth. Between 1945 and 1970, though the supply area hardly expanded, water supply had increased more than fivefold from six to 32 million cubic meters per year and the number of connections had risen threefold from 48,000 to 159,000.

2.5   Synthesis of the Chapter

This Chapter introduced the WSS sector, as it forms the sector studied in the thesis at hand. A case of service expansion in the Dutch province of Friesland was examined to illustrate the traditionally public character of the WSS sector. This case, which described service expansion in that province between 1922 and 1970, highlighted the role played by the municipal governments in the province of Friesland. In fact, in almost all cases, public service providers, which have improved services significantly, have been able to do so because of support from the political realm, which extended well beyond the activities of making policy. Without active political support and government guarantees, few WSS providers would be able to 'turn around' performance.

The Chapter identified that policy makers and practitioners in the WSS sector are confronted with major challenges, due to the increasing demand world-wide for WSS services. A central role in shaping the institutions to respond to these challenges is to be played by government through a reform process defining the new institutional context. Policy makers contemplating about alternative institutional arrangements need to manoeuvre carefully in view of the externalities associated with WSS services provision.

This thesis is an effort to gain additional insight in the consequences of implementing institutional changes in the WSS sector. The need for additional insight is particularly relevant for the WSS sector in view of the ambiguities in characterizing the nature of WSS services. These ambiguities appear throughout the discussions and analyses of the value of institutional change in the WSS sector. On one hand WSS services are often perceived as a public good, but applying the theoretical definitions, it surfaces that it is really a private good. Also WSS services are often labelled as typical monopoly goods; while a closer look at the WSS services process several parts have definite competitive elements. Third, WSS services are perceived to have merit good characteristics, but since the Dublin principles the economic nature of WSS services has been more pronounced. Policy makers attempt to find a way in shaping the institutions in which WSS providers need to operate within the sometimes internally conflicting characteristics ascribed to WSS services.

Section I   Introductory Section

Chapter 3 Introducing Neo-liberalism

This Chapter introduces neo-liberalism. An explorative case in Cochabamba, Bolivia, is investigated to surface relevant elements of the research topic[8].

3.1   Introduction

In this Chapter a major development in the WSS sector is singled out, e.g. the neo-liberal institutional changes. Neo-liberalism is defined and its theoretical grounding is explored. In particular the theoretical basis of neo-liberalism is held against the sector specificities of the WSS to understand its applicability. Next, it is identified how neo-liberalism manifests itself in the WSS sector. Then, an overview is provided of scholarly inquiries into the effects of neo-liberal institutional changes in the WSS sector. The Chapter ends with an illustrative case study that highlights the complexities for policy makers to adopt the neo-liberal agenda.

3.2   Neo-liberalism

The term of 'neo-liberalism' is used to denote a group of neoclassical-influenced economic theories, right-wing libertarian political philosophies, and political rhetoric that portrayed government control over the economy as inefficient, corrupt or otherwise undesirable (Hart-Landsberg, 2006). Although neo-liberalism is not a unified economic theory or political philosophy, it can be observed that broadly all neo-liberal reform measures find their rationale in neoclassical economics. Neo-liberalism proposes that human well-being can best be advanced by liberating individual entrepreneurial freedoms and skills within an institutional framework characterized by strong private property rights, free markets and free trade (Harvey, 2005). Distinct streams of thought have dealt with the implications of neo-liberal reform measures (Villalonga, 2000):

1. Welfare economics.
2. Contract theory.
3. Property rights theory.
4. Principal-Agent theory.
5. Public Choice theory.

A closer look at the theoretical grounding reveals that in applying them to the WSS sector there is some level of ambiguity: welfare economics ascribe numerous benefits to competition, but the extent to which competition in the WSS sector is limited compared to other sectors; contract theory ascribes numerous benefits to contracting, but contracts closed in the WSS sector are often incomplete. And for the property

---

[8] Parts of this Chapter (including the Cochabamba case) have been published in Schouten, M. and K. Schwartz, 2006. *Water as a Political Good: Implications for Investments.* In: International Environmental Agreements: Politics, Law and Economics. Vol. 6. No. 4; pp. 407-421. Springer

rights theory, which suggests the benefits of private and transferable ownership, the picture becomes blurred due to the monopoly status that also a private owner will have as a WSS operator. With respect to the supporting notions from Public Choice theories, it may not be feasible to exclude the WSS sector from governmental interferences. Governments and public officials cannot escape their responsibility in the eyes of the general public (their voters) for any malfunctions in WSS provision, also if the service provision has been delegated to another (private) party. Hence it might be that the WSS sector has some unique characteristics that reduce the relevancy of the neo-liberal theories. As Okten and Arin (2006: 1539) state:

> However, ambiguity of the theoretical literature about ownership in less competitive markets seems better justified since the empirical literature is also less conclusive on the effects of ownership in such markets.

The following sections will address the supporting theories and will analyse the degree of relevancy of these theories to the WSS sector.

3.2.1   Welfare economics

The central element in liberalization is the introduction of *competition*. The reasons to introduce competition are several, and most of them are based on the economical dogma of a perfect competitive market structure where suppliers have an overwhelming incentive (Glaister, 1996) to achieve maximum overall efficiency in order to sustain and deliver a value for money product or services that customers choose to buy from a range of products. The benefits of competition form one of the roots of economics as a science. Adam Smith's book on The Wealth of Nations, dating from 1776, helped to create the modern academic discipline of economics and provided one of the best-known intellectual rationales for free trade, capitalism and libertarianism. It suggests that under specific conditions, market mechanisms will yield accurate incentives and foster efficient resource use. The main advantages ascribed to competition are (i) consumer sovereignty, (ii) the optimum allocation of resources (iii) the absence of expensive bureaucracy and administration, and (iv) the motivational influence of free enterprise (Vickers, 1995). However, analysing these four benefits, it can be concluded that they are only partly relevant to the WSS sector.

Consumer sovereignty relates to the ability of consumers to persuade producers to produce more of a particular product by making that product more profitable (i.e. by offering a higher price for it). This ability maximises consumer welfare as each individual consumer can decide his demand for a certain product and the price he is willing to pay for it. The combined effect of all consumers that exercise this ability maximises the consumer welfare. Unfortunately in the WSS sector only in rare cases consumers have the possibility to select alternative suppliers. Hence, the essential element of competition is often lacking (Van Dijk, 2003).

The benefit that competition induces an optimum allocation of resources in society pertains to the argument that consumers determine the value of goods by their willingness to pay for the good. From a society point of view a good becomes more valuable if more consumers are willing to buy it. Since such good is valuable for the society it is important that more scarce resources are made available to produce this good. So consumers are determining through their preferences and buying behaviour (their effective demand) the allocation of scarce resources in society. In a situation when there would be competition in the WSS sector, supply of water services will be determined by effective demand (thus income for the WSS provider). Hence, the part of the population that is able to pay for the service provision will steer the service provision and the allocation of the scarce resources. Such may result in perverse consequences, like a great inequality in service provision by serving only those that can indeed afford to pay the bills.

Also the third benefit of the absence of expensive bureaucracy is only partly valid in the WSS sector, due to the many market imperfections inherent to the sector. In all cases extensive regulation is to be set in place to secure the public interest related with WSS provision. Also in the case of competition, a certain degree of bureaucracy is inevitable for the WSS sector.

The fourth benefit related to the motivational influence of free enterprise seems also to hold little relevance in the WSS sector. The supporters of the market system argue that the possibility of making profits under the free market system will act as an incentive to individuals to start up businesses and supply what customers desire (Blokland *et al.*, 1999). A severe limitation in the WSS sector related to this argument is that the resources of land, labour and capital are immobile. Prospective suppliers are not able to easily expand or contract in response to changing consumer demand due to high capital intensity of the sector.

3.2.2   Contract theory

Contract theory studies how economic actors can and do construct contractual arrangements, generally in the presence of asymmetric information. Contract has become a fundamental metaphor for the neo-liberal changes in the WSS sector (Walsh, 1995). More and more, a WSS provider is operating in a "nexus of contracts", rather than a bureaucratic hierarchy.

Contract theory ascribes numerous benefits to the use of contracts (Hart, 2003). A contracted party will pay more attention to human resource development and draws the best out of their employees in terms of productivity, welfare and creativity. The contracted party will be able to access additional funds and make better use of available funds. Also contracts may specify the use of cutting edge technological innovation and research. And entrusting WSS provision by contract to private parties may lead to more transparency, since contract will specify detailed performance indicators and reporting and monitoring mechanisms. A last benefit of contracting is that the contracted party may be more "attuned" to the customer satisfaction, from quality and service control to reliability and rapid expansion of services to the consumers (Njiru and Sansom, 2001).

However, the potential to realize these ascribed benefits in the WSS sector is relatively small. Rivalry for getting a delegation contract is often muted or absent, either because governments find negotiated contracts more convenient or because bidders engage in collusive behaviour. Also the long duration of delegation contracts in the WSS sector constitutes another source of uncertainty and risk. Bidders must be able to eliminate uncertainty and make risk manageable over sometimes very long contracting periods (duration of concession contracts is often 30 years). When outcome is difficult to link to activities, as in WSS provision, contracting is a problem. A last difficulty is that both the government as the contractor should be able to terminate the contract without suffering major repercussions. Contractors will be reluctant to terminate the contract and write off sunk costs. Governments can ill afford to terminate a contract when this will have serious repercussion in the public domain (Braadbaart, 2005).

3.2.3   Property rights theory

According to the property rights theory a fundamental distinction between private and public enterprises concerns the transferability of property rights (Braadbaart, 2005). The existence of a market to transfer property rights offers capital gains to potential owners who can conceive of efficient operating procedures.

The attenuation of ownership rights in public enterprises has a direct link to managerial behaviour. Managers of public WSS providers have greater opportunity to increase their own welfare at the expense of the employer's wealth. Moreover, since direct benefits can only be internalised by a public official during his tenure in office and the costs of his decisions beyond his political horizon do not affect his net worth, managers of public WSS providers are expected to have a higher rate of time preference than private providers. This argument implies that input choices in public service providers will tend to be biased away from long-term capital investment and towards utilization of labour or other variable factors (Crain and Zardkoohi, 1978).

Evidence from WSS providers that have been privatised suggests that the property rights theory is relevant. Sawkins (2001) has examined whether the transfer in ownership through take-overs and mergers in the UK water industry has resulted in better performances of the water providers. His research showed that already the threat of a takeover by another company is enough to push management to increase the efficiency. To provide an example, he gives an illustration of the situation of South West Water.

> A leading example of this in England and Wales was South West Water's announcement of a £ 15 customer rebate and 20.4 percent increase in interim dividends during its 1996 defence of hostile take-over bids from Severn Trent and Wessex Water. In the end, the bids were blocked by the Secretary of State for Trade and Industry. The threat of take-over, however, was enough to induce South West to put its management under considerable pressure to make good the reduction in profit that resulted from the customer rebate.

An interesting, very recent, development in this respect is that financial investors are also becoming involved as owners in the WSS sector. Currently, three of the ten water & sewerage companies in the UK are owned by private financial investors. For example, the water infrastructure in London is since 2006 owned by a consortium led by Macquarie Bank, Australia's largest securities firm[9].

3.2.4   Principal-Agent theory

The before discussed Property rights theory has a link to Principal-Agent theory. Determining the preferred owner raises the basic problem of Agency[10]. The central idea is simple: a *Principal* wants to induce another entity, the *Agent*, to perform a given task, which is associated with a cost to the Agent. The Agency problem relates to the difficulty of ensuring that the Principal is faithfully served and that the Agent is fairly compensated. Donahue (1989) makes a basic distinction in this regard: a private

---

[9] This consortium bought Thames Water from the German utility RWE for £8 billion. Macquarie Bank acquired 11 percent of the utility, with the rest held by the Macquarie European Infrastructure Fund, Macquarie European Infrastructure Fund II and other investors (Pinsent Masons, 2007).
[10] The word Agency, confusingly, holds several meanings. Here it refers to a type of relationship, not a governmental office.

party, in exchange for a price, agrees to deliver a product; while a civil servant, in exchange for a wage, agrees to accept instructions. In general, the Agent's interest do not entirely coincide with those of the Principal; the Principal does not have complete control over the Agent; the Agent has only partial information on the Principal's interests; and the Principal has only partial information about the Agent's behaviour. The Agency relationship consists in the reliance of a Principal upon an Agent with an agenda of its own. The Principal-Agent theory emphasizes the importance of the interaction between the public authorities responsible for service provision and the public or private bodies that are executing the service provision[11]. In the context of the Principal-Agent theory, one could state that traditionally the local public authorities acted both as Principal and as Agent.

Where ownership and managerial control over a company have become separated, it is often difficult to get the managers to work on behalf of the owners, instead of letting their self-interest prevail (De Wit and Meyer, 1994). Principals in the WSS sector are mostly local and national governments assuming the responsibility for service provision and safeguarding the public interest. These Principals rely on Agents to perform the management of the service provision. Agents in the WSS sector take many shapes; they can be municipal departments operating very closely to the Principal, but they can also be operating at arms' length distance from the Principal in the form of public limited companies or private contractors. Principals set up the legal and operational limitations for the Agent to operate in, and try to induce Agents to act in the interest of the Principal's objectives. According to the Principal-Agent Theory, Agents are tempted to serve their own interests, even when it is to the detriment of the Principal. The Principal needs to condition her actions on some information that is privately known by the Agent. The Principal could simply ask the Agent for the information, but Agents may not report it truthfully unless the Principal gives them an incentive to do so, either by monetary payments or with some other instrument that the Principal controls. Since providing these incentives is costly, the Principal faces a trade off that often results in an inefficient allocation.

The Principal is assumed to choose the mechanism that maximises the expected utility, as opposed to using a particular mechanism for historical, political or institutional reasons. As such the link with liberalisation in the WSS sector is obvious. Principals, in the shape of governments, need to assess in what way they optimally can arrange the linkage with an Agent to maximise the public interest it targets. In some countries Principals choose to have an Agent very close by, so it is able to directly influence its operations. If it chooses the arrangement as such, the Principal needs to realise there is a trade off. It might be easier for the Principal to influence the Agent's behaviour, but it negatively affects competition and the entrepreneurial spirit with its benefits as efficiency, innovation and customer orientation. On the other hand, if the Principal chooses to create a large degree of separation with the Agent there is also a trade off: it might motivate competition and the entrepreneurial spirit, but the Principal needs to put in place a costly regulatory system to still be able to (indirectly) influence the Agent.

---

[11] Also Game Theory (Fudenberg and Tirole, 1991) is supporting the Principal-Agent Theory. In Game Theory the WSS sector is treated in a special class of games of incomplete information known as games of (static) mechanism design. Examples of these games include monopolistic price discrimination, optimal taxation, the design of auctions, and mechanisms for the provision of public goods.

The Principal-Agent theory is relevant to the WSS sector. It can be observed that the roles of Principals and Agents are changing and so do the relations between them. Many institutional arrangements with their respective forms of ownership, organization and governance are capable of meeting the needs of customers, the community and the environment in a successful manor.

3.2.5   *Public Choice theory*

The Public Choice School's central argument is that politicians pursue their own utility rather than the public interest. Accordingly, they impose goals on public WSS providers that can lead them to gain votes but can conflict with efficiency. For the general public, who are the ultimate owners of the WSS provider, the costs of monitoring their behaviour (e.g. information gathering, lobbying) are likely to offset the benefits (e.g. less taxes, or more efficient public spending). This is not the case, however, for interest groups such as trade unions, which makes state-owned enterprises an easy target for rent seeking activity.

In the WSS sector it is common use that governments directly manage the service provision, relying on their hierarchical relationship with the provider. Current estimates place the share of the world's population that is being serviced by private service providers at no more than 3%-10%, of which the majority resides in high-income countries (OECD, 2003). However, the governmental involvement in the WSS sector is often considered to be one of the main causes of inadequate service provision. According to Haarmeyer and Mody (1998), providers which are directly managed by the government, are infamous for being inherently inefficient, overstaffed, manipulated by politicians to serve short-term ends, unresponsive to consumer demands, and – particularly in low-income settings – inclined to provide subsidized services to urban middle class and leave the urban and rural poor unserved. Also Foster (1996) finds that a system in which a government is both controlling and executing the WSS provision is institutionally deficient. In such system the classic poacher-gamekeeper problem comes up, which results in a confusion of regulatory and operational roles of the government. Next, according to Foster (1996), there is a tendency of public WSS providers to base day-to-day decisions on political rather than technical or commercial criteria as a result of external influences from government. Also Spiller and Savedoff (1999: 2) share in this respect Foster's analysis. Basing themselves on their research of the provision of WSS services in Latin America, they find that inadequate provision of WSS services is largely a consequence of:

> the nature of the sector, coupled with a nation's political institutions, which together create incentives for government-owners of public utilities to behave opportunistically, for the service providers to operate inefficiently, and for the consumers to withhold support from the sector. As such, the water services sector under these circumstances has a tendency toward a low-level equilibrium from which it is difficult to escape.

Hence, an often-proposed solution for improving service is to separate service provision from policymaking and regulatory functions. This separation of service provision from policy making may be done by private sector involvement or by so-called 'agencification' (Schwartz and Schouten, 2007). In the case of private sector involvement the responsibilities for service provision are (either entirely or partly) delegated to the private sector by way of contract. In the case of agencification, service provision is separated from policy making by creating (public) autonomous agencies, which are to operate at arms length of the government owners. The suggested benefits of this separation of functions are multiple. First of all, attributing different functions to different organizations allows for a system of checks and balances, in which different organizations monitor the functioning of other organizations. This essentially addresses the earlier-mentioned poacher-gamekeeper problem discussed by Foster (1996). Another benefit of such a separation of functions is that it would reduce political interference in day-to-day operations (World Bank, 2004). The public WSS provider would then concentrate on the day-to-day management of service provision, whilst the political realm would concentrate on the policy-framework within which service provision takes place. A third benefit of such a separation is that it would allow for greater clarity and accountability for the various organizations executing the different functions (World Bank, 2004).

Despite the ascribed benefits to limiting the political influence on the service provision, it needs to be noted that the idea that a stringent separation of the political realm from the management of the service provider is possible seems unrealistic, if not naive (Schwartz and Schouten, 2007). Especially consumers leave it "difficult [for the politicians] to escape being called to account for the performance of services" (Deakin and Walsh 1996: 35). Even if the tasks of policy formulation, service provision and regulation were allocated to separate agencies, politicians would still be held accountable by the general public with respect to the actual services provided. These citizens would demand action if they consider the quality of services provided to be inadequate. This accountability draws the politicians back into the actual provision process of WSS services. Pollitt and Bouckaert (2000: 146-147) echo this sentiment when they find that:

> [A]ny suggestion that public management can be radically depoliticized [...] is either a misunderstanding or flies in the face of evidence from many countries. The allocation of say, health care resources or decisions about educational standards or major public infrastructure projects are all inherently 'political' decisions, whether they are taken by powerful politicians or tough public managers [...]. The public will often see the political authority as ultimately responsible – or at least sharing responsibility – however many ministers may protest that these are technical or professional decisions which have been taken by the appropriate officials.

Since the public interest is at stake in the delivery of WSS services, government plays an important role in shaping the right institutional context for good performance. Brown (2002: 126), in arguing why privatisation does not lead to a separation of politics from service provision, provides the following observation:

> It is absurd to think that a private [operator] will sink a large amount of capital into an enterprise and then unilaterally disarm himself politically. Obviously investors will use all legal means, including political, to protect their interests. Similarly, it is unreal to expect that social expectations will terminate or diminish merely because [responsibilities] have been transferred to the [private sector]. It seems obvious that investors will seek to manipulate a system to their benefit, and equally obvious that politicians, interest groups, advocacy organizations, and others will continue to push for their own objectives.

In short, the realities of the WSS sector are such that in most countries governmental involvement is "a fact of modern life for water suppliers" (Scharfenaker, 1992: 26).

Realities of WSS services are much more complex with all stakeholders turning to the political realm to further their interests. Moreover, at times, an active role of policy makers in the service provision process may even benefit the level of services provided. In other words, the role played by the political realm is more complex than is suggested by those calling for the stringent separation of functions.

3.3   Manifestations of neo-liberalism in the WSS sector

Neo-liberalism manifests itself in the WSS sector most notably through liberalisation, privatisation, and Private Sector Involvement (PSI)[12]. The three terms are often called in one breath due to the numerous commonalities and shared benefits. As Budds and McGranahan (2003: 6) stated:

> While terms as "private sector participation", "privatisation" and "Public-Private Partnerships" (PPP) are in common use, there is sufficient ambiguity to justify noting some of the different ways in which they are used in literature.

Memon and Butler (2003a) make a useful distinction in this regard. They distinguish between 'privatization-in-full' from 'privatization-in-part'. Privatization-in-full refers to divestiture arrangements, entailing the transfer of assets ownership (including infrastructure) and management from the public sector to the private sector. In privatisation-in-full essentially the only shift is that the previously public supplier of the service is transformed into a private supplier, since the underlying idea of privatisation is that the government is regarded as less able to provide WSS services compared to private parties. Divestiture arrangements are markedly different from liberalisation and from PSI. In these type of arrangements the essence is the ownership shift from public to private, which not necessarily has to be combined with competition or a sense of partnership.

Privatization-in-part is referred to as a Private Sector Involvement (PSI) type of arrangement. PSI is a generic term describing the relationship formed between the private sector and public bodies often with the aim of introducing private sector resources and/or expertise in order to help provide and deliver public sector assets and services. The central element in a PSI is the (often formal) partnership that the actors conclude with one another. In many cases such arrangement is shaped in the form of a contract (Van Dijk, 2003). In a PSI the relation between the private supplier and the public buyer is central. The essence of PSI is the establishment of the partnership, and again such can be achieved without any competition or shift in ownership. For example in the WSS sector, the management of services is delegated to private operators through concessions or lease contracts to private operators, but the water supply system remains publicly owned (French model). PSIs describe a wide variety of loose, informal and strategic partnerships, to design build finance and operate (DBFO) type service contracts and formal joint venture companies. The broad application of PSI is a distinct strength as it reflects the wide diversity of practice that has developed to facilitate private sector participation in the provision of public infrastructure.

---

[12]Bakker (2003) attempted to overcome the difficulties with the 'old' neo-liberal terms, as privatization and liberalisation, by introducing two new terms, e.g. "marketization" and "commercialisation". Commercialisation is sometimes also referred to as "corporatization". Marketization refers to signify the introduction of the logic of the market into water resources management and/or water supply. In this respect it includes both "privatisation" (the shift in ownerships and control from the public to private companies with private capital) and "commercialisation" (a reworking of water management institutions along commercial lines but not necessarily with private sector involvement).

The term of 'liberalisation' is separated from the terms privatisation and PSI due to its primary focus on competition. Liberalization is defined as 'a process by which competition is introduced in situations or sectors hitherto characterized by exclusive or special rights or monopoly granted to historical operators' (Van Dijk, 2003). In this respect liberalisation often goes hand in hand with a removal (or modification) of rules in a particular market (deregulation or re-regulation). In liberalisation the effort is targeted at increasing the level of competition between the suppliers of the good. A competitive market is a market with at least several sellers, which should bring down the prices and allows the buyer some choice (Van Dijk, 2003). The guiding theme of liberalisation is the increase in competition, and such may very well go without any private party involved; let alone any private ownership or partnership. In the one-and-a-half decade of heightened attention to liberalising the WSS sector, several types of competition have been found or rediscovered. Four main possibilities exist in order to implement effectively rivalry into network industries:

1. Product-Market competition (competition *in* the market). Competition *in* the market is when the buyers may directly and at any time choose the suppliers of the good they purchase.
2. Franchise bidding (competition *for* the market). Competition *for* the market occurs when potential (public or private) operators bid competitively for a temporary delegation contract, otherwise known as a franchise contract. This approach, as proposed by Harold Demsetz (1968), is widely used in the WSS sector, and is sometimes combined with unbundling of activities.
3. Quasi competition (comparative competition, yard-stick competition or benchmarking competition). An alternative form of competition is to compare performance of different companies operating on different geographical areas but on similar services. The comparisons can be made for segments of the utilities' operations and can cover a range of variables such as capital maintenance costs, operating costs, prices, quality of service, etc.
4. Self supply. A user may decide to get himself the water. He may construct his own network within a legal basis for his own final use. For instance, this is the case of a firm that needs a large quantity of water but the connection to the public network is more costly than self-supplying.

Further clarifying the distinctions between privatisation, liberalisation and PSI use can be made of the concept of markets. Markets are created when exchange takes place; as a market is a collection of buyers and sellers that transact a particular product or product class (Kotler, 1998). This economic definition of a 'market' is very close to the traditional interpretation of the village square which is once a week transformed into a market square where buyers and sellers meet to exchange their goods (Lipsey *et al.*, 1987).

Figure 3 presents how the three main neo-liberal terms can be distinguished using the concept of markets.

Figure 3 Neo-liberal manifestations

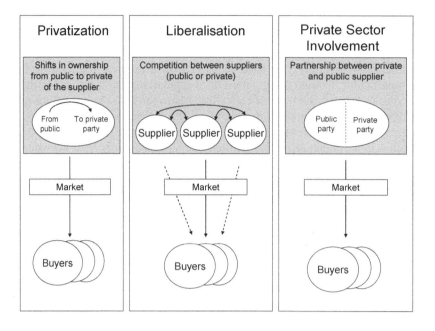

It follows that only in the case of liberalisation (the middle block in Figure 3) the opportunity is created (or strengthened) for buyers to select from more than one supplier. In the other two cases (privatisation and PSI), such is not necessarily so. Privatisation and PSI relate merely to a change at the level of the supplier, either through a shift in ownership, or in a sharing of tasks and/or responsibilities. Hence, liberalisation can take place without any private sector involvement or privatisation, and vice versa. It is also possible to carry out these three processes at the same time or one after the other. Apparently competition (liberalisation) shifts in ownership (privatisation), and private sector involvement (PSI) are complementary to another. Each of them has its own set of advantages and disadvantages. The concepts can be isolated or effectively combined in tailored neo-liberal reform measures, keeping a close eye on the net effect they have on the whole of the sector. There is some evidence from several studies that competition is generally more important than ownership, per se, in explaining performance improvements in the WSS sector (e.g. Wallsten, 2001; Zhang *et al.*, 2003a, 2003b). Sachs *et al.* (2000) for example, examine the empirical evidence across 24 change economies and conclude that ownership alone is not enough to generate economic performance improvements. Also Galal *et al.* (1994) demonstrate that while ownerships matters, competitive markets reinforce the benefits of private ownership. The impact of private sector involvement might in this regard be smaller in the WSS sector compared to the other infrastructures as the potential for competition in the WSS sector is much lower than in sectors like electricity, gas and telecom.

3.4   The neo-liberal trend in the WSS sector

Since the 1990s one can observe within the WSS sector a trend of governments to change the existing institutions by giving more importance to private sector involvement and the use of market mechanisms. Before, governments were in many cases acting as both the responsible as well as the executing entity of service provision, but such has changed significantly in the last two decades. For the WSS sector, the early 1990s proved to be a turning point as several events led to an increased attention to neo-liberalism; i.e. the privatisation of the WSS sector in the UK in 1989, the 1992 Dublin Statement of water as an economic good, and the changed course of the World Bank embracing the ideas of the Washington Consensus. These events were part of, and accelerated the first wave of neo-liberalism (Smith, 2004). This first wave is characterized as a 'roll-back the state' form of neo-liberalism (Peck and Tickell, 2002). WSS service provision was being decentralized and devolved to private parties. Many governments pursued a policy of private sector participation through partnerships, more decentralized management, an emphasis on demand-based provision and a greater degree of cost recovery (Smith, 2004). However, in the late 1990s this first wave was transformed into a second wave of neo-liberalism due to growing arguments against private sector involvement in WSS service provision. This second wave is characterized as a 'roll-out the state' form of neo-liberalism (Peck and Tickell, 2002). In this second wave emphasis was put on the roles of the government, as a precondition to reap the benefits of the market mechanism. Regulatory mechanisms and New Public Management (NPM) entailed new forms of governance that introduce different relationships between the government and the market. In NPM government is mimicking the private sector in providing the services (Schwartz, 2006). Remaining constant in the WSS delivery debates during both the first and the second wave is the dominant view that the private market logic is a more efficient method for delivering public services.

Numerous statistics show the impact of the neo-liberal reform agenda on the WSS sector. One of the main references in this respect is the voluminous Pinsent Masons Yearbook, which is annually published by a London based law firm. The yearbooks monitor the trend of increasing private sector involvement in the WSS sector since 1987. In this respect they use a set of variables, i.e. population served by the private sector, frequency of contract awards, and average size of contract awards (separated by type of projects), number of projects with private sector involvement (also distinguished by type). It needs to be noted that the statistics only partly succeed in indicating the extent to which indeed more use is made within the WSS sector of the market mechanism. Nevertheless, data throughout the years show that for all of indicators there is upward trend; hence it can be rather safely assumed that –to a certain extent- the neo-liberal reform agenda has been embraced by the global WSS sector. According to the calculations of Pinsent Masons (2008) the private sector involvement continues to increase in the WSS sector. In 1999, 5% of the world's population was served to some extent by the private sector. Since, 2006 this has increased to 10% of the world's population and to 11% in 2007 and 2008, with between 731 and 751 million people served. There were some 272 contracts with private parties in 1999 against 935 in 2008 (see Graph 1 below).

Graph 1 New PSP contracts awarded in the WSS sector

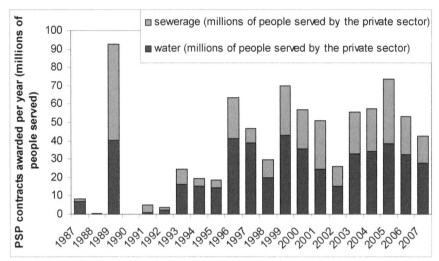

Source: Pinsent Masons (2008).

The propagation of the use of private parties and the market mechanism fits within the neo-liberal reform agenda. McCarthy and Prudham (2004: 275) indicated the profound influence of neo-liberalism on institutions with the following statement:

> Neo-liberalism is the most powerful ideological and political project in global governance to arise in the wake of Keynesianism, a status conveyed by triumphalist phrases as "the Washington consensus" and "the end of history".

The origins of neo-liberalism are complex, yet one focal point is that neo-liberalism is a reaction against Keynesianism during the 1970s (McCarthy and Prudham, 2004). Many governments after World War II adopted the recommendations of Keynes for state intervention into markets. However, in the 1970s, it was perceived that these state interventions were often counterproductive. Markets are complex, and governmental interventions may destabilize the market. This thinking led to a new liberal movement, called neo-liberalism. Neo-liberalism rose to prominence in the USA and the UK during the 1980s under President Reagan and Prime Minister Thatcher. It also influenced international development thinking through the wide-scale adoption of the so-called Washington consensus. Williamson (1990) invented the term 'Washington Consensus' to refer to the lowest common denominator of policy advice being addressed by the Washington-based institutions (like the World Bank and the IMF) to Latin American countries as of 1989.

Voigt and Engerer (2002) observe that New Institutional Economics (NIE) has been largely neglected when the neo-liberal thinking was developed. They state (127-128):

> ... advocates of the New Institutional Economics have not been at the forefront of giving policy advice. A decade into change in Central and Eastern Europe, the Washington consensus seems to have lost much of its attraction. It has been observed that growth and development in countries that have implemented macroeconomic policies in line with the Washington consensus display considerable variation in their

economic performance. It has been hypothesized that variation can be explained by differences in the institutional quality of the countries. Policy advice might thus soon be sought from institutional economists.

The argument made in the above-presented statement of Voigt and Engerer that New Institutional Economics theory may be valuable (for policy makers) to provide additional understanding of the implications of neo-liberal reforms, captures the main relevancy of the thesis at hand. Neo-liberalism should in the interpretation of the thesis primarily been viewed as a change in institutions, affecting the conduct of the executing agencies and subsequently the services (performance) delivered to the users.

3.5   Empirical enquiries on neo-liberal changes in the WSS sector

In this section, an analysis is made of the available literature on the implications of institutional context (particularly private versus public) on the WSS sector. The analysis contributes to a better understanding in directly relating institutional context to performance.

Since about two decades numerous studies have been executed to establish concrete evidence on the relation between neo-liberal institutional changes and the performance of WSS providers. Crain and Zardkoohi (1978) refer even back to debates in 1850 over nationalizing certain British enterprises. In other sectors research has produced "mounting evidence" (Jerome, 2004: 1) that associates neo-liberal institutional changes with improved performance. The mounting evidence, Jerome is referring to, is based on studies executed by several authors in other sectors (for example: Boubakri and Cosset, 1998, Megginson and Netter, 2001; Dewenter and Malatesta, 2001). However, other authors contest Jerome's notion of superior private performance (Seppala *et al.*, 2001).

The empirical literature can be classified into two groups: cross sectional studies, and longitudinal studies (Villalonga, 2000). Table 1 provides an overview of a selection of the empirical enquiries. The availability and variety of longitudinal studies is sharply in contrast with the cross sectional studies.

Table 1 Selection of empirical studies

Study	Type	Sample	Outcome
Mann and Mikesell, 1976	Cross sectional	USA	Public more efficient
Morgan, 1977	Cross sectional	USA	Private more efficient
Crain and Zardkoohi, 1978	Cross sectional	24 private - 88 public (USA, 1970 data set)	Private more efficient
Bruggink, 1982	Cross sectional	USA	Public more efficient
Feigenbaum and Teeples, 1983	Cross sectional	27 private - 262 public (USA, 1970 data set)	No significant difference
Boland, 1983	Cross sectional	USA	Private more efficient
Teeples, Feigenbaum and Glyer, 1986	Cross sectional	USA	No significant difference
Byrnes, Grosskopf and Hayes, 1986	Cross sectional	59 private – 68 public (USA, 1976 data set)	No significant difference

Study	Type	Sample	Outcome
Fox and Hofler, 1986	Cross sectional	20 private – 156 public (USA, 1981 data set)	No significant difference
Teeples and Glyer, 1987	Cross sectional	USA	No significant difference
Byrnes, 1991	Cross sectional	USA	No significant difference
Lynk, 1993	Cross sectional	UK	Private more efficient
Lambert, Dichey and Raffiee, 1993	Cross sectional	USA	Public more efficient
Raffiee *et al.*, 1993	Cross sectional	USA	Public more efficient
Bhattacharyya, Parker and Raffiee, 1994	Cross sectional	32 private – 225 public (USA, 1992 data set)	Public more efficient
Bhattacharyya, Harris, Narayanan and Raffiee, 1995	Cross sectional	USA	Private more efficient
Shaoul, 1997	Cross sectional	UK	No statistical difference
Menard and Saussier, 2000	Cross sectional	France	No statistical difference
Saal and Parker, 2001	Longitudinal	UK, 1985-1990 data set, and 1990-1999 data set	No improvement in total productivity growth since privatisation (1989)
Estache and Rossi, 2002	Cross sectional	22 private - 28 public (Asia, ADB data set)	No significant difference
Estache and Kouassi, 2002	Cross sectional	3 private – 18 public (Africa)	Private more efficient
Clarke and Wallsten, 2002	Cross sectional	Africa	Private higher coverage
Kirkpatrick, Parker and Zhang, 2004	Cross sectional	9 private - 101 public (Africa)	Private more efficient
Faria, da Silva Souza and Moreira, 2005	Cross sectional	13 private - 135 public (Brazil, SNIS 2002 data set)	Private marginally more efficient
OfWat International Comparator Reports from 1996 to 2007	Cross sectional	England & Wales, Australia, Netherlands, Portugal, Scotland, USA, Canada	No significant difference

It is to be noted that the original contention of the OfWat International Comparator studies, mentioned at the bottom of Table 1, is not to assess whether privately owned companies outperform public WSS providers. Instead the studies are executed to set out 'the relative performance of the England & Wales industry in world terms by making international comparisons in a number of key performance areas, from unit costs to leakage rates' (OfWat, 2007: 4). OfWat has produced annual international comparator reports since 1996. In the first editions a comparison was made only with Sydney Water from Australia, but the most recent edition from 2007 includes comparisons with a group of utilities from Scotland, Scandinavia, the USA, Northern Ireland, the Netherlands, Portugal, Canada, and Australia. OfWat has selected six main indicators to compare the performance. However, OfWat notes that they ran into problems once trying to compare the performance ratios from different countries.

An analysis of the results presented in Table 2, shows that the performance of the English and Welsh appears comparable to the performances of water providers in other countries.

Table 2 Key performance comparators of OfWat

Performance indicator	Comparison problems and approach of OfWat	Conclusion for the 2007 report
1. Bills to customers	A simple comparison of the bill levels across various countries cannot take into account the level of service to customers, investment to maintain and enhance assets or the presence of subsidies. However, comparison of bill levels can provide a useful high-level appraisal.	Customer bills in England & Wales are comparable with those in other countries.
2. Customer service levels	Information comparing levels of customer's service is generally difficult to find.	The level of customer service that companies in England & Wales offer appears to be high compared to other countries.
3. Water quality and environmental performance	Comparable water quality information is difficult to find for many countries. Consequently OfWat only considers variations in quality for the water industries in England & Wales, Scotland and Northern Ireland.	Companies in England & Wales generally achieved better drinking water compliance than their counterparts in Scotland and Northern Ireland.
4. Water delivered, leakage and water efficiency	Comparison of volumetric data that accounts for all types of water use should always be treated with some caution. Although the measurement of the volume of water put into distribution systems is relatively straightforward different definitions for water balance components such as distribution losses, supply pipe leakage and customer use can make direct comparison difficult.	Water use in England & Wales lies in the middle of the range found in Europe. Reported water leakage in the Netherlands is extremely low. There is evidence to suggest that differences in the operating environment can explain some of the disparity between the countries considered.
5. Unit costs and relative efficiency	Assessed via the main components of water service unit costs per property and per unit of water delivered. The former may be the more useful measure, as levels of water use vary widely between countries and volume may be a poor determinant of costs for some areas.	Comparisons based on the number of properties suggest that total costs are broadly similar in the Netherlands and England & Wales. Comparison of volumetric unit costs suggests that the costs in the Netherlands are higher than in England & Wales. The English and Welsh companies appear more cost efficient than Scottish Water.
6. Network activity	Data is collected on capital expenditure and network performance to indicate the expenditure of maintaining an enhancing service delivery, and the effect this has on service to	Following a program of investment, the condition of Scottish Water's water mains and sewer pipes are seen to be improving.

Performance indicator	Comparison problems and approach of OfWat	Conclusion for the 2007 report
	customers.	
7. Financial performance.	The financial performance of an organization is important whatever its regulatory framework. Some key financial ratios provide an indication of a company's ability to finance its functions.	The levels of return on the capital base are broadly similar in all countries. Financial indicators across countries appear to be broadly similar to England & Wales, though there is significant variation between individual companies in each country. Accounting practices and assumption between countries are likely to distort the ability to make direct comparisons.

Source: based OfWat (2007).

OfWat considers the data used in the report are sufficiently comparable to allow them to draw generalised conclusions. However, OfWat also warns that the data should not be based for specific regulatory or business decisions on the comparisons in the studies. OfWat acknowledges the limitations in using the data. According to OfWat, the approaches to define, collect, and use performance indicator information vary dramatically between the WSS providers in the sample. Therefore they recommend that interpretation should be done with care (OfWat, 2007: 6).

> Data is not very robust and can only be used to put the regulated companies' performance in an appropriate wider context ... Data is not fully comparable, but exposes differences that challenge the current performance levels of the regulated.

The information is specifically intended to challenge and reflect, not with the intention to provide any conclusive evidence of the (superior) performance of a neo-liberal institutional arrangement to a traditional public institutional arrangement.

Analysing Table 1 and Table 2, it is clear that the empirical evidence is less robust than one would hope for, both in depth of the analyses as in uniformity of outcomes (Megginson and Netter, 2001; Renzetti and Dupont, 2004). Sometimes the private sector scores better, other times the public sector outscores the private sector. Vickers and Yarrow (1991: 117) reviewing empirical studies of performance differences between public and private operators with natural monopoly elements, like water, concluded:

> the results of empirical studies are very mixed: some give the advantage to public ownership, others to private ownership, and yet others can find no significant difference between the two. Substantial performance differences among utilities do, nevertheless, exist, both within and between countries.

Martin and Parker (1997: 93) share the notion of Vickers and Yarrow, basing themselves on a survey of empirical studies on the effects of neo-liberal changes. They state that:

> on balance it seems that neither private nor public sector production is *inherently* or *necessarily* more efficient. (emphasis of authors)

It is remarkable to notice that reform measures as introducing competition and privatisation are undertaken without clear empirical evidence that they are beneficial to the service provision (Brown and Iverson, 2004). Obviously and undoubtedly the institutional neo-liberal changes incorporate substantial costs. Such cost arise in the form of lengthy negotiations, hiring consultants and lawyers, establishing new organisations like regulatory bodies, etcetera. These costs are burdened in an article of faith on the positive implications of the institutional changes. Still, governments and opinion leaders embrace the reform measures believing that they will bring a solution to the perceived inefficient and politicised service provision of public operators. In some cases these characteristics of underperformance have a ring of truth, but such views risk becoming a stereotype on which many policy decisions are premised. Accurate or not, these stereotypes often helped to justify reform measures that were, in fact, driven by shifts in political priorities (Brown, 2002).

According to Neal *et al.* (1996) there are five potential drivers that brought the wide scale adoption of the neo-liberal reform agenda in the WSS sector, dramatically changing the views on how network utilities should be owned, organized and regulated in both the advanced industrial economies as the developing and change countries. These drivers are:

- Societal. The belief that the use of the market mechanism can help to satisfy unmet basic water needs.
- Commercial. The belief that more business is better.
- Financial. The belief that the private sector can mobilise capital faster and cheaper than the public sector.
- Ideological. The belief that smaller government is better.
- Pragmatic. The belief that competent, efficient water-system operations require private participation.

Interestingly, all neo-liberal drivers are according to Neal *et al.* (1996) based on a belief system. As McCloskey (1998) has argued (neo-liberal) reform measures are to a large extent decided on rhetoric, not on facts. Such would coincide with the view that in most countries, policies driven by commercialism, ideologies and pragmatism are central to the WSS sector transformation (Gleick *et al.*, 2002). The belief systems of policy makers towards the advantageous implications of neo-liberal reform measures in the WSS sector is apparently to some extent biased, often borrowing arguments from other sectors.

In this respect Green (2000) divides policy makers in two kinds of people: Panglossians and Pragmatists[13]. Panglossians generally think that water needs to be managed as a purely private good through markets and economic instruments. They believe that once some market failures are corrected, the best possible of all worlds will result. The opposing view is presented by the Pragmatists; who argue that economic efficiency is not the only objective to be satisfied. In their view, economic efficiency is only one of many policy objectives. In short, Panglossians believe; Pragmatists have questions. Overall the Pragmatist's responses to the favoured

---

[13] Savenije (2002) also distinguished two schools of thought that seem very similar to the Panglossian and Pragmatist distinction. The first group (similar to the Panglossian group) maintains that water should be priced at its economic value; the market will then assure that the water is allocated to its best uses. By and large, this point of view is advocated by the World Bank. The second school (similar to the Pragmatist group) thinks that decision on the allocation and use of water should be based on a multi-sectoral, multi-interest and multi-objective analysis in a broad societal context, involving social, economic, environmental and ethic concerns.

solution offered by the Panglossians for problems the WSS sector is coping with, are summarised in Table 3:

Table 3 Panglossians versus Pragmatists

Solutions proposed by Panglossians	Responses of Pragmatists
The objective for society is to have an efficient provision of WSS services.	Society has multiple and internally conflicting goals with respect to WSS provision.
Perfect markets do and can exist in the WSS sector.	The perfect market is an idealised construction. Instead one should focus on the problem of governance.
Prices are the only effective incentive to changing water consumption behaviour.	Next to prices, there are other ways to encourage or penalise particular behaviour.
Universal water metering is an obvious requirement in the WSS sector.	Meters don't save water; it is what people do in response that saves water.
Privatisation is the obvious answer in the WSS sector.	There are no obvious answers: common property and municipal companies have historically been the way WSS provision is managed.
Green taxes are the way to deal with water pollution.	It is more important what you do with the money collected from the green tax charges.
The way to deal with over-abstraction is to create tradable water abstraction permits.	You need to start with a system of law that recognises hydrological realities.

Source: Green (2000), modified by author.

The first answer of Panglossians to any question is that some form of market should be created. Panglossians hold the belief that perfectly, competitive markets can exist and should be pursued. Conversely, Pragmatists believe that where co-operation will be against the public interest, competition is necessary to prevent producers from exploiting consumers. Pragmatists argue that markets are inherently unstable because the incentive for all producers is to seek a monopoly or failing that to collude with other producers in an oligopoly. Panglossians believe that a market will always do better than a government and that governments are unfortunate necessities whose role should be limited as far as possible. Pragmatists are only interested in markets to the extent to which they enable competition where competition should be expected to drive down prices and drive up the quality of the service provided.

(Panglossian) arguments about the superiority of private sector compared to the public sector in managing WSS services built the case to engage the private sector. For example, in the UK just three years before the privatisation of the WSS providers, the White Paper on Water Privatisation (CMnd 9734, 1986: paragraph 38) concluded:

> Private enterprise is both more flexible and readier to pursue energetic and innovative approaches than the public sector. The demands of the market will give management and staff the impetus they need to secure greater efficiency. Freeing the authorities from the constraints imposed by state ownership will help them to carry out their tasks with vigour and imagination.

Contributions from the research community in providing further insight and guidance on the applicability and use of the neo-liberal reform agenda for the WSS sector are urgently demanded. To further explore the main relevance of the theme to the thesis, below a case study is revisited of one of the best known and most discussed cases of failed private sector involvement in the WSS sector (Schouten and Schwartz, 2006).

3.6   The landmark case of Cochabamba

This case is the 40-year concession contract, including a large investment component, in the city of Cochabamba, Bolivia. The contract, which was awarded to a private consortium headed by International Water Limited[14], was terminated less than six months after the contract was signed. The period of social struggle, which preceded the cancellation of the contract has since become known as the 'Cochabamba Water War'. The case of Cochabamba is useful in the sense that it highlights both the drivers for adopting the neo-liberal reform agenda, as the difficulty for policy makers to beforehand estimate the impact of the institutional change. The case clearly shows that the applicability of neo-liberal institutional changes has a strong dependency on the local social, economic and cultural context and developments.

Prior to 1985 the collapse of the international market of minerals, which represented about 40% of Bolivian exports (World Bank, 1996), led to a period of continued macro-economic instability. This economic crisis was characterized by periods of hyperinflation, reaching a level of 11,000% - 25,000% in the year 1985, and culminated in tremendous government debt. As a reaction, the government implemented an ambitious stabilization and neo-liberal structural adjustment program in 1985. Although the reforms introduced from 1985 onwards did manage to achieve a measure of macro-economic stability, economic growth rates remained low. In 2000, per capita GDP was about US$ 1,000, making it one of the poorest countries in Latin America. Moreover, the country went into a recession in the late 1990's with per capita GDP declining by 1.9% in 1999 and 'staying flat' in 2000 (Hilderbrand, 2002). Poverty was not significantly reduced (Finnegan, 2002) with 67% of the population living below the poverty line (Hilderbrand, 2002) and the population of Bolivia became increasingly opposed to the prevailing neo-liberal economic strategy (Nickson and Vargas, 2002). "By late 1999 this growing economic crisis was giving rise to widespread protests in many parts of the country, spearheaded by teachers and police demanding pay rises". In the case of Cochabamba there was the additional factor of coca-leaf farmers who were opposed to the US-financed programme to eradicate coca-leaf farming. "For many Bolivians, the new Law and concession contract together symbolized all that was wrong with the neo-liberal development strategy" (Nickson and Vargas, 2002: 139).

The reforms focused on liberalization of markets, developing an export orientation, liberalization of interest rates, deregulation of labour markets and fiscal adjustment. A key part of this strategy was to increase the role of the private sector in sectors such as oil and gas, telecommunications and transport (Camacho, 2005). The role of the private sector was also to be expanded in the WSS sector. In 1997 the first water concession contract was granted to Aguas de Illimani, a consortium headed by SUEZ-Lyonaisse des Eaux, in the city of La Paz-El Alto. The Bolivian government viewed the 30-year contract as a central piece of their strategy to improve service provision in the La Paz-El Alto area (Komives, 1999). Following the La Paz-El Alto concession contract the government initiated a similar process for the city of Cochabamba. Hence, in 1999 a concession contract was signed for service provision for the city of Cochabamba and surrounding areas with a private company called Aguas de Tunari (AdT). Although it seemed logical to continue with the strategy of engaging the

---

[14] International Water Limited is jointly owned by the US construction company Bechtel and the Italian energy company Edison (Lobina, 2000).

private sector through concession contracts following the successful start of the La Paz-El Alto concession, the WSS sector in the city of Cochabamba showed significant differences from the sector in La Paz-El Alto.

First of all, service coverage in the city of Cochabamba only reached 57%, meaning that 43% of the population had to rely on other sources, such as community service providers, private wells and private vendors for their water. Due to the poor service coverage of the public water utility, the *Servicio Autónomo Municipal de Agua Potable y Alcantarillado de Chochabamba* (SEMAPA), the city of Cochabamba was characterized by a large number of alternative service providers[15], who provide services where the 'formal' (public) utility fails to do so. These alternative providers were small-scale private operators, essentially running their business on commercial principles or community based systems, operating more as a community cooperative. The prominent role of the alternative systems is important in the context of the rights attributed to AdT in terms of the concession contract. The concession contract granted AdT exclusive rights for the provision of WSS services. Moreover Law 2029 stipulated that "concessionaires would have exclusive rights over the concession area" (Assies, 2003:17). The reason for granting such a right is that exclusivity of service provision reduces the revenue risk to which the private operator is exposed. Potential customers have, short of moving out of the service area, no option but to buy water from the private operator. This essentially guarantees a certain level of income for the private operator. The problem, however, was that in Cochabamba city many alternative service providers were operating, and under the contract these systems would be forced into contracts with the concessionaire. The community service providers had little production costs and as such could provide cheaper services (though perhaps not services of better quality) to the community that established them and had financed the required infrastructure. The idea of having to pay money to AdT for the service provided by the community system was not appealing to the community. The private service providers, operating on commercial principles obviously viewed the exclusivity rights as a threat to their business. "Hence, the concession contract threatened the vested interests of other provider groups in the concession area" (Nickson and Vargas, 2002: 142). These vested interests not only pertained to the community and private service providers but also to the companies supporting the operation of these organizations, such as private drilling companies which drilled wells for these systems (Nickson and Vargas, 2002).

Second, the area was suffering from severe water scarcity. This water scarcity was caused by increased use of water resources, less rainfall in the Cochabamba area and the reduction of the water retention area in the Tunari National Park (Camacho, 2005). Water scarcity also had significant consequences in the city of Cochabamba and the surrounding areas, which together make up the Central Valley. Already before the implementation of the concession contract fierce competition for water resources existed in the Central Valley between agricultural users and urban water users and between Cochabamba and the neighbouring province of Quillacollo. Small farmers in the four municipalities surrounding the city of Cochabamba used groundwater resources for irrigating their land. With growing urban demand the public water utility SEMAPA had drilled groundwater wells as well, leading to a situation of rapidly dropping groundwater levels, despite assurances from SEMAPA that the wells would

---

[15]   It is estimated that the figure for alternative service providers for the city of La Paz is approximately 7.7% and for El Alto 3.8% (Camacho, 2005), indicating that the role of alternative service providers is much more prominent in Cochabamba than in La Paz-El Alto.

in no way affect water levels (Assies, 2003). The concession contract (supported by Law 2029) established that in order to fulfil the requirements of service expansion AdT would be allowed to identify and develop future water resources. The small farmers thought that the concession contract and Law 2029 represented a "threat to their established rights and that water for irrigation would henceforth be charged, despite assurances to the contrary" (Nickson and Vargas, 2002). Not surprisingly, the contract was immediately opposed by local organizations such as the Cochabamba Department Federation of Irrigators' Organizations (FEDECOR) (Assies, 2003).

In order to address the issue of water scarcity, the Cochabamba contract, at the beginning of the tendering procedure, was to include the development of the Misicuni Multipurpose Project (MMP)[16]. Feasibility studies carried out by the World Bank, however, led to the conclusion that due to the high investment costs, estimated at US$ 300 million, the MMP was not feasible and an alternative, less costly project, was proposed. This alternative was the Corani project, which was estimated at US$ 90 million, and would be executed by Corani S.A., a private hydroelectric company that generates electricity in the Cochabamba Valley (Nickson and Vargas, 2002). As such, the original MMP was replaced by the Corani project. Before the concession process for the Corani project was completed, however, the municipality of Cochabamba challenged the proposed Corani concession on the grounds of non-compliance with procurement law[17]. The challenge was successful and the concession process for the Corani project was aborted (Nickson and Vargas, 2002). Following the cancellation of the tendering process for the Corani project, the Ministry of Investment initiated a second tendering procedure. This time the companies would have to include the MMP option in their bids. However, only one consortium, AdT entered a bid and failing to meet the standard of three competing bids, the tendering procedure was declared invalid. Following the invalidation of the tendering process negotiations ensued between the government and AdT concerning the concession and the MMP option (Nickson and Vargas, 2002). These negotiations were successfully completed in June of 1999 and AdT was awarded a 40-year concession (including the MMP). Important to note, however, is that when the concession contract was signed in September 1999, there was no legal framework, which actually supported the concession contract. Law 2029 on Drinking Water and Sewerage was adopted two months after the signing of the concession contract, essentially legalizing the earlier signed contract with AdT (Assies, 2003).

The concession system introduced by Law 2029, as mentioned, foresaw that AdT would take control of the hundreds of alternative systems in the city of Cochabamba and would install water meters on those systems. This prospect and the rumours concerning possible tariff increases led organizations (such as the Civic Committee and the Federation of Neighbourhood Associations) that had supported or even co-signed the contract to become more critical of the concession. The initial opposition to the AdT contract came from the FEDECOR and the College of Engineers. Soon a Committee for the Defense of Water and the Popular Economy was established "with engineers Osvaldo Pareja, Gonzalo Maldano and Jorge Alvarado among its driving

---

[16] The Misicuni Project involved the construction of a US$130m dam, 4,000 metres above sea level, a hydroelectric power station and a US$70m, 20 km long tunnel to transport water from the Misicuni River through a mountain to the valley of Cochabamba.

[17] Although non-compliance with procurement law was the official reason for challenging the Corani project "it is widely believed that the real reason for the appeal was because the mayor supported the MMP option". This support "reflected pressure from politically influential Bolivian engineering and construction companies, who expected lucrative contracts from the MMP" (Nickson and Vargas, 2002:135). Or as Finnegan (2002) explains: "some of his main financial backers stood to gain from the Misicuni's Dam's construction".

forces" (Assies, 2003: 22). This Committee organized a forum as early as July 1999 to express its criticism of the Aguas del Tunari concession. Those who, a month before, had agreed to the contract largely ignored the forum. As such, the initial protests "remained a matter of professional organizations and certain environmentalists without a broad social base" (Assies, 2003: 22). The initial opposition by FEDECOR received a further stimulance by the cooperation between FEDECOR and the Departmental Federation of Factory Workers of Cochabamba (FDTFC). With the cooperation of the FDTFC and FEDECOR the Defense Committee was transformed into the Coordination for the Defense of Water and Life. The '*Coordinadaro*' was to become the most pro-active of the groups protesting the concession and included a wide range of professional organizations. "The new coalition introduced a rural-urban dimension and brought a significant broadening and radicalisation to the committee, which had mainly appealed to professional sectors and some environmental groups" (Assies, 2003: 24). Assies (2003: 25) provides an interesting example of this broad base when he describes how five hundred workers from the Manaco shoe factory went to the city centre "to protest the layoff of 60 workers and to call for continued action on the water issue". As such, even if the interest groups had their own specific issues, which they protested, the 'water issue' became a shared issue. With the interests of the farmers and the alternative service providers threatened whilst the concession contract, symbolic of the neo-liberal policies, was awarded to a consortium headed by a foreign company (even worse, by an American company in which the former American Secretary of State, George Schultz was a Director), the 'water issue' became the 'shared issue' around which the interests groups rallied. Berg and Holt (2002: 3) note that "the example of Cochabamba illustrates that the Bolivian government's policy objective to improve and expand the city's water and sewerage networks did not adequately consider the concerns of some affected parties". Although this observation is correct, it obscures the fact that the 'water issue' not only became important to those directly affected by it, but it grew to be a symbol of opposition against the government and fifteen years of neo-liberal economic policies.

From the Cochabamba case several observations can be made. First, after 15 years of structural reforms and 'neo-liberal' economic policies the political and economic environment was such that the opposition of the Bolivian population to the government's policies was growing. Essentially, the population increasingly opposed 'neo-liberal' policies, which emphasized the private, economic good dimension of (infrastructure) services, without seeing much in return. Second, the characteristics of the WSS sector and the government policies in the Central Valley were such that the WSS sector made a highly suitable sector around which to organize protests. For example, the fierce competition in the 1990s between different water users had forced the various interest groups, such as the farmers, to familiarize themselves with the debate in relation to the water legislation (Assies, 2003). Moreover, a large portion of the population in Cochabamba was serviced by alternative systems, of which the small-scale providers had a strong stake in the developments in the WSS sector in Cochabamba. Against this backdrop the government implemented the concession contract and Law 2029, which essentially allowed the monopolization not only of the service provision systems, but even of the water resources. This led to immediate opposition by the aforementioned groups, which were quick to protest both the contract and Law 2029. Soon after, however, with the cooperation between the factory workers union and farmers organization FEDECOR the opposition really broadened its base.

3.7   Synthesis of the Chapter

In this Chapter, the worldwide adoption of the neo-liberal agenda by policy makers for the WSS sector was identified as the major institutional change. In view of the trend data presented in the Chapter it becomes clear that many policy makers since the 1990s have embraced the neo-liberal agenda. The Cochabamba case study is instrumental in surfacing the conflicting perceptions on WSS services. The explorative case description from Cochabamba serves as a landmark case on the lack of understanding of the value and effects of neo-liberal reform measures. It emphasizes the urgent need to create insight into possible consequences and enabling conditions for neo-liberal institutional changes.

The neo-liberal institutional changes in the WSS sector are part of larger worldwide development in other sectors. The WSS sector was even quite late in adopting the neo-liberal principles. In the beginning of the 1990s the first wave of neo-liberal institutional changes can be identified, which is characterized by a replacement of governments by private parties ('roll-back the state'). In the late 1990s, the first wave was transferred into a second wave in which the government instead of retreating from the sector aimed to redefine its role ('roll-out the state'). One can distinguish three forms how neo-liberal changes manifest themselves in the WSS sector: privatization-in-full (divestiture), privatization-in-part (PSI), and liberalization. The three forms can take place at the same time, or one after the other.

Investigating the motivations to introduce neo-liberal changes, it appears that these are mostly based on a belief system; not on any compelling scientific evidence. Green (2000) distinguishes in this respect people with a firm belief in the use of market mechanisms (so-called Panglossians), and people that believe that every situation demands its own solution (so-called Pragmatists). In many cases the influences of the Panglossians in introducing neo-liberal reform measures can be recognized. Also the case of Cochabamba illustrates such since this concession contract was part of a larger reform effort relevant for many sectors. There was a definite need in the area to both increase the total amount of water services to be provided, as well as to the number of people. On the initiative of the government institutional changes were implemented aiming to respond to the increase in demand. Of direct influence to the initiative in Cochabamba was the perception of success in starting a concession contract in La Paz-El Alto. Policy makers seemed unaware of the need to tailor the institutional change to the local and historical context in Cochabamba in terms of scope and timing. Essentially they were hoping for the best, without having any idea on the possible implications of the institutions they implemented. Despite the risk of going into unknown waters, neo-liberal institutional change was implemented.

The Chapter identifies that some of the theories underpinning neo-liberalism may only be partly applicable to the WSS sector. As was described earlier, the WSS sector has several characteristics that may hinder the relevancy of theories and practices in other sectors. Welfare economics, contract theory, Agency theory, Property Rights theory and Public Choice theory are all analysed for their relevancy to the WSS sector, and it shows that in many instances their applicability is limited. Only the Agency and Property Rights theories seem to hold a fairly strong argument for neo-liberal institutional changes. On the one hand the low relevancy of neo-liberal theories increases the relevancy of the research at hand since it raises questions about how they work out in the WSS sector. On the other hand it also complicates the research since it becomes hazardous to ascribe possible changes in conduct to the institutional change to the foundations of the neo-liberal agenda, e.g. competition and private sector involvement.

The Chapter clearly demonstrates the inability of researchers to generate conclusive evidence on the implications of adapting the neo-liberal reform agenda in the WSS sector. Normative and anecdotal discussions on the implications of neo-liberal reforms continue to outpace systematic investigations. Several reasons are identified which may have caused the inability of researchers to provide such evidence. Such is supported by Shirley and Walsh (2001) who find that both the theoretical literature and the empirical literature are not conclusive about the merits of private ownerships in monopolistic markets. This finding establishes the relevancy of the thesis at hand.

In sum, this thesis is concerned with the governmentally motivated institutional changes in the WSS sector, and more specifically with the adoption of the neo-liberal agenda. The main conclusion that can be made of this introductory section is that there is a clear need for further insight in the implications of neo-liberal institutional change in the WSS sector.

Section II: Research Design

Chapter 4:   **Research Objective and Approach**
Chapter 5:   **Analytical Framework**

Section II. Research Design

Chapter 4 Research Objective and Approach

This Chapter outlines the research objective and approach of the thesis at hand. It introduces the construct of 'strategy' as an intermediate variable in the research design.

4.1   Introduction

The Research Design section builds on the Introductory Section by defining how the research at hand is intended to contribute to the existing body of knowledge on the value of neo-liberal institutional changes in the WSS sector. Research design refers to the basic plan of the research and the logic behind it, which will make it possible to draw more general conclusions from it. The research problem is made researchable by setting up the study in a way that it will produce specific answers to specific questions (Oppenheim, 1992).

The Research Design section is composed of two chapters; the first chapter outlines the general research approach, the second chapter provides the details in operationalizing the research. A lot of attention is provided in this Section to the inclusion of 'strategy' in the research design, which is a novelty compared to the majority of studies executed in the WSS sector. The function of 'strategy' as an intermediate variable is an important choice within the context of the thesis and requires interpretation and rationale. Hence, due attention is paid in this chapter to better understand the construct of 'strategy' and the main theoretical notions from strategic management.

4.2   Research objective

The previous three Chapters, composing the Introductory Section, have identified neo-liberal changes as the dominant contemporary development in the WSS sector. Given the profound influence of neo-liberalism on the WSS sector, it is remarkable to note the existing ambiguity with respect to its value. Scholars have not been able to give conclusive evidence on the merits or harms of neo-liberalism for the WSS sector. Some empirical investigations point to the direction of superior private performance, other studies find the public party to be the better one, and sometimes there was no difference found between the performances of the private and public parties. Consequently, policy and decision makers in the WSS sector are implementing neo-liberal institutional changes merely based on a (Panglossian) belief system, not supported by a proper insight into its possible consequences. In this respect both from a societal as from a scientific point of view it is important...

... to increase the understanding of the value of neo-liberal institutional changes in the WSS sector.

The above statement is what the research at hand is aiming for. The research objective is essentially similar to that of the empirical studies identified in Chapter 3 in Table 1 and 2. However, the research design is deferring from the existing body of literature by adopting an alternative approach. The following section will elaborate on the alternative approach in achieving the above-mentioned research objective.

4.3   Introducing 'strategy' as the intermediate variable

The key feature of the research approach adopted in this thesis is to make a comprehensive analysis of the relations between the (shifting) institutional context, the (changed) conduct of the WSS providers and the (change in) performance. Martin and Parker (1997: 170) support such approach by stating:

> ... in so far as ownership and competition are important, they impact on performance through *an internal adjustment process*" (their emphasis).

This statement of Martin and Parker captures the main idea behind the thesis. It is one of the main propositions of the thesis that WSS providers are purposive institutions and management of WSS provision is a creative and proactive process. Institutional change may have an influence on the 'internal adjustment processes' of the WSS provider, which then subsequently may change the level of service, the tariff, the investments, and other performance elements. It cannot be taken for granted that the conduct of private WSS providers is indeed different from that of a public WSS provider. In this respect, the thesis touches "one of the most fundamental problems in strategy scholarship, namely the nature of causation" (De Rond and Thiéart, 2007: 2). Strategy scholars are confronted by the question whether performance is determined by particular resource configurations, by competitive or industry dynamics, by institutional pressures, or by other elements outside of the organization's control. Two polar approaches are available in organisational science to address this question: determinism versus voluntarism.

The majority of the body of literature explaining implications of neo-liberalism are based on a deterministic foundation. They assume that institutional changes have an independent impact on performance; therefore failing to elaborate on the source and strength of this impact (Ward *et al.*, 1995). The assumption to ignore 'conduct' in earlier research design is shared with (and possibly influenced by) mainstream approaches from the traditional economic tradition, like the Structure-Conduct-Performance (SCP) paradigm, the Contingency theory and the Organizational Ecology approach. The SCP paradigm, the early Contingency theory and the Population Ecology perspective assume that performance of organizations is to a great extent determined by the environment, excluding largely the role of managers (Brown and Iverson, 2004; Jeffery, 1990). Although these theories are primarily applied to competitive sectors, it seems logical also for researchers in the WSS sector to emphasize the importance of the external environment (especially the regulatory environment). The SCP paradigm, the early contingency theory and the Population Ecology theory are shortly described in the following sections in order to understand better the rationale for the traditional research approach linking institutions directly to performance in existing empirical enquiries.

The SCP paradigm is one of the building stones of the Industrial Organisation theory. The essence of the SCP paradigm is that market structure largely drives the market performance (Aldrich, 1979; Hannan and Freeman, 1977). It is to be noted that market structure differs from institutions. The SCP paradigm interprets market structure in terms of the number of buyers and sellers in the market, the growth of the market, the existence of substitutes, the cost structures, product differentiation, and entry barriers to the market. The joint market conduct then determines the collective performance of the organizations in a given market (Bain, 1968; Mason, 1957). According to SCP theorists, the market structures primarily determine the performance, and the change in conduct of individual organisations can be assumed as an automatism. Market conduct is less emphasised in the literature on SCP, and in empirical enquiries even neglected. In numerous studies scholars linked performance with the market structure characteristics, statistically (cor)relating market structural characteristics to performance (Porter, 1981). According to Porter (1981: 611):

> A final crucial aspect of the Bain/Mason paradigm [e.g. the SCP paradigm] was the view that because structure determined conduct (strategy), which in turn determined performance, we could ignore conduct and look directly at industry structure in trying to explain performance. Conduct merely reflected the environment.

In a similar vein, proponents of the Organizational Ecology perspective contend that the success or failure of organizations is determined by inertial and environmental forces. In their view, organizational survival is largely dependent on environmental selection (Hannan and Freeman, 1977, Aldrich, 1979). One significant premise underlies thinking in organizational ecology is that processes of change in organizational populations parallel processes of change in biotic populations. Alterations in organizational populations are, in their interpretation, largely due to demographic processes of organizational foundings (births) and dissolutions (deaths). Thus in the view of organisational ecologists, the role of managers becomes less important to the survival (performance) of its organisation, while the institutional context is the dominant determinant. Organisational ecologists argue that (Singh and Lumsden, 1990: 163):

> the evolution of populations of organisational forms can best be studied by examining how social and environmental conditions influence the rates at which new organisations are created, the rates at which existing organisations die out, and the rates at which organisations change forms.

The unit of analysis of the SCP paradigm and the Organizational Ecology Perspective is at industry (or population) level, and therefore emphasizing the interaction between individual actors. In this sense the SCP paradigm is particularly useful for determining the likely average performance of an industry, but less useful for sorting out the different performances of individual organizations. This is also supported by the notion of Bush and Sinclair (1991) that studies of an industry as a whole may miss important intra-industry strategic differences, and company-level analyses may not generalize to the industry level. The early contingency theory is in this respect different from the SCP paradigm and the Population Ecology theory, although it also emphasizes the deterministic role of the environment. While the SCP paradigm and Population Ecology theory target the dynamics at the level of an entire market, the contingency theory is focussing also the individual organisation operating in a market. Traditional contingency theorists (Child, 1972) see the conduct of organisations as a necessary response to the environment, more than as influencers of the environment. As Miller states (1988: 281):

> They [e.g. the traditional contingency theorists] have taken performance as a function of the match between organisation and environment, without considering strategy.

Miller interprets conduct of organisations in the above statement through the construct of 'strategy'. Also in the thesis at hand a choice is made to proxy conduct through the construct of 'strategies'. This is an important choice implicating the research design to a great extent.

By definition the large majority of strategy scholars reject absolute determinism by the external environment of performances of organisations, and tend to embrace the approach of voluntarism. After all, it is the assumption of managerial discretion that gives rise to such strategy process orientations at the design, planning, positioning, entrepreneurial and cognitive schools (Mintzberg *et al.,* 1998). Also it is one of the assumptions in this thesis that individual organisations can partly determine their performances by altering its conduct. In this light it is valuable to increase further insight into the three-step relation of 'institutions–conduct–performance' of WSS providers. Porter (1981) supports a framework as the one proposed in the thesis. He claims that no automatism between changes in structure and changes in conduct should be taken for granted, just as well as no automatism between changes in conduct should be assumed with changes in performance. This coincides with the view that organizational factors and their fit with the environment are the major determinants of performance of an organisation (Hansen and Wernerfelt, 1989). Managers in this perspective play a crucial role in determining the performance, much more than the institutional context (or industry structure) in which they operate. Advocates of the Strategic Choice paradigm (Child, 1972) suggest that managers are the main arbiters of organization direction and performance. They suggest that since managers choose the domain of the firm's activities, decide on its resource allocation priorities, and design competitive manoeuvres, they are directly responsible for its performance.

4.4   The relevancy of the strategic management field to the WSS sector

Several developments point into the direction that the field of strategic management will gain importance in the WSS sector. One is that the WSS sector seems to be become more and more similar to competitive sectors due to increased private sector involvement and competition. In this respect, the results of a study conducted by Sisson (1992) are noteworthy. He found a clear relation between the perceived level of competition of managers of WSS providers and the extent to which strategic planning practices were implemented. Sisson (1992) showed through a survey among 248 water providers that WSS providers that perceived higher levels of competition are more likely to have implemented a formal strategic planning process. Almost 85% of the WSS providers that perceived their business environment as 'very competitive' stated to have a strategic planning process, compared to only 50% of organisations that perceived their environment as 'not-at-all competitive'.

Another development of the increased importance of the strategic management field for the WSS sector is the ongoing trend in the field of strategic management to focus efforts on other sectors than competitive ones (Llewellyn and Tappin, 2003). Ring and

Perry (1985) argue that the application of the principles of strategic management developed from private sector studies could help public sector organizations despite the fact that strategic management research is virtually non existent in public and other not-for-profit organizational contexts. Nevertheless, it needs to be noted that since Ring and Perry's writing a lot has changed, also in the WSS sector. Contemporary strategic management literature is moving away from the competitive element in defining strategy, and puts more emphasis on the creation of value for customers and other stakeholders. The concept of strategy as a weapon is increasingly replaced by strategy as a means to create value for any organisation, either public or private. An example of such is the recent strategy definition of Kaplan and Norton (2004) as "the way an organization describes how it intends to create value for its shareholders, customers and citizens".

Illustrative of the increased importance of strategy management literature for the WSS sector is the evolution the Dutch water sector went through in using strategic planning tools (see Box 1). Just like the study from Sisson (1992), the Dutch case on the historical development of strategic planning tools shows the relevancy of strategic management to the WSS sector. A possible reason for the evolution in the use of strategy formulation tools might be that the WSS providers increasingly perceive the external environment as becoming more turbulent. Also the evolution may be accounted to a greater awareness and acceptance among managers of WSS providers to engage into alternative ways to undertake planning, being influenced by developments from other sectors.

Box 1 Historical development of strategic planning in the Netherlands

Since the birth of the Dutch drinking water sector in 1851 when King Willem II gave permission to establish the first Dutch drinking water company in Amsterdam (Klostermann, 2003), the character of its' policies and strategies is typified through long time horizons. These long time horizons can be explained by the long lifetime of capital investments and the value given to sustainable service delivery. If it was judged that support was needed for long range planning decisions, water companies relied on extrapolating past trends to provide an understanding of the future. The past proofed to be a solid basis for future planning as the levels of uncertainties and change were marginal. As the water sector can be characterised by robust developments, for a long time traditional forecasting has prospered (Becker and Van Doorn, 1987). In support of long term planning, the construction of models was and is frequently done to calculate future water and investment demands, possibly also since the sector is dominated by an engineering culture infected by modelling exercises from more technical disciplines.

Making use of the future methodology tree as developed by Coyle (1997), one might say that the Dutch drinking water made dominantly use of passive, defensive approaches or the more active analytical anticipatory approaches to support long range planning decisions. The passive, defensive and analytical anticipatory approaches were (and possibly still are) perfectly appropriate and respectable as the internal and external environment for the drinking water sector was relatively stable for a long time. The geographical area was predetermined and the monopoly position of each service provider was and currently still is untouched in view of the impracticality of introducing competition in the market. Data about customers was available, as well as the demographical increase and water uses.

Next, in the field of technological innovations not much changed since 1851, although treatment technologies became more sophisticated. But still the production and distribution process is simple and comparable to 150 years ago. Also, the product is still the same as 150 years ago although the quality obviously increased. The main uncertainties that the sector is dealing with, are coming from governmental interferences. Changes in government provide turbulence into the sector. Sometimes legislation for the sector became tighter or less rigid, and different financial mechanisms were put in place to make the sector more efficient or effective. But even from the government side the public status of the water companies was never really challenged in view of the service of the public interest.

Just since the turn of the century, an interesting development starts taking place when a new and more active anticipatory technique was introduced in the Dutch water sector to support strategic decision-making in the form of multiple scenario analysis. In the year 2000 the first initiatives were established at sectoral level to develop multiple scenario analyses. Within these projects some feasible future end states of the sectors were defined, including a dynamic sequence of interacting events, conditions and changes that were necessary to reach that end state

The following multiple scenario building projects were undertaken in the Dutch drinking water sector recently. The most influential and widely known project is probably 'De Kartonnen Doos"- project that was carried out in 2002 by a project team of representatives from water companies, supported by the branch organizations VEWIN and Kiwa, and a consultancy agency CIBIT. This project was succeeded by a project called 'Horizon scanning' that currently serves to scan the developments with respect to the identified key trends (Van Eekeren, 2002). Also consultancy agencies conducted recently some futures research for the Dutch water sector. One is the project 'Water Voorzien', dating from 2001, executed by the management consultancy agency Twijnstra & Gudde to contribute to the discussion about the water services in 2015 in the Netherlands. Another is the multi scenario analysis executed by another consultancy agency Arthur D. Little. Also the scientific community added a multi scenario description through the Euromarket project from 2002 to 2005. This project financed by the European Commission aimed at identifying scenarios at pan-European level and member state level in view of possible liberalisation. Another more specifically aimed project was the AWWARF IMS project in 1999/2000. On the basis of research at six locations (three in the Netherlands, three in the USA) scenarios were developed on water treatment. The project was executed in cooperation with the University of Central Florida.

4.5   Defining strategies

There is no lack of definitions of strategy. The definition from Andrews (1987: 18-19) has traditionally been considered as one of the more complete definitions:

> The pattern of decisions in a company that determines and reveals its objectives, purposes, or goals, produces the principal policies and plans for achieving these goals, and defines the range of business the company is to pursue, the kind of economic and human organisation it is or intends to be, and the nature of the economic and non-economic contributions it intends to make to its shareholders, employees, customers, and communities.

Mintzberg (1987a) pointed out the concept of strategy is not consensual in nature and has significantly evolved during the past decades. Mintzberg identified a total number of five definitions of strategy, which in his opinion are all applicable depending on the context in which strategies are used. The following five definitions are according to Mintzberg all relevant:

1. *Strategy as a plan (intended)* – a direction, a guide or course of action into the future, a path to get from here to there. Strategy provides overall direction to the whole enterprise, and as such is assumed to have critical influence on the success or failure of a company. One of the early authors on strategy, Chandler (1962: 16), provides a similar definition, e.g. "the determination of the basic long-term goals and objectives of an enterprise, and the adoption of courses of action and the allocation of resources necessary for carrying out these goals".

2. *Strategy as a pattern (realized)* – a consistency or behaviour over time. The notion of strategy as a pattern of actions is close to the previously quoted definition from Andrews (1987) and the selected definition for the thesis from Croteau *et al.* (1999).

3. *Strategy as a position* – namely locating particular products in particular markets. One objective of an overall strategy is to put the organization into a position to carry out its mission effectively and efficiently. An organization's strategy must be appropriate for its resources, circumstances, and objectives. The process of strategic management involves matching the companies' strategic advantages to the environment the organization faces. This interpretation is shared by Porter (1996), when he argues that strategy is about competitive position, about differentiating yourself in the eyes of the customer, about adding value through a mix of activities different from those used by competitors[18].

4. *Strategy as a perspective* – namely an organization's fundamental way of doing things. Strategies integrate an organization's goals, policies, and action sequences (tactics) into a cohesive whole. Changing position within perspective may be easy; changing perspective, even while trying to maintain position, is not. A definition provided by Wood (1999) fits this type of definition. He describes strategy as "the process by which an organization generates, develops and maintains a robust business design capable of both exploiting its current distinctive capabilities (its fitness function) on or near its current fitness peak and exploring its strategic landscape and business ecosystem for entrepreneurial opportunities beyond the lifecycle of its current business design (its sustainability function) away from its current peak".

5. *Strategy as a ploy* – that is a specific "manoeuvre" intended to outwit an opponent or competitor.

The above five definitions from Minzberg are indicative for the disagreements between strategy scholars. Despite the volume of literature, or because of it, there is no common agreement on what strategy is (Hart, 1992). Strategy management literature is vast and, since 1980, has been growing at an astonishing rate. Mintzberg *et al.* (1998) identified over the years a total of 10 different strategy schools of thought. Each school of thought has its own merits and contextual dependency. In view of the notion of multiple definitions of strategy that are relevant for different uses, a choice needs to be made within the thesis on how strategy is interpreted. One particularly appropriate within the context of the thesis is the one from Croteau *et al.* (1999). Croteau *et al.* (1999: 2) define strategy as:

---

[18] It should be noted that Porter writes about competitive strategy, not about strategy in general.

the outcomes of decisions made to guide an organization with respect to the environment, structure and processes that influence its organizational performance.

This definition captures the most relevant elements from the longer and more comprehensive definitions, like the one from Andrews (1987), and makes the connection with performance.

4.6   Dimensions of strategy

De Wit and Meyer (1994) suggest that there are three dimensions of strategy: strategy context, strategy process, and strategy content.

- Strategy context concerns the *where* of strategy that is in which organization and in which environment are the strategy process and the strategy context embedded.
- The strategy process concerns with *how* the strategy is established, *who* is involved and when do the necessary activities take place.
- Strategy content is according to De Wit and Meyer concerned with the *what* of strategy: or the pattern of actions through which organizations proposes to achieve desired goals, modify current circumstances and/or realize latent opportunities.

Boyne and Walker (2004) note that strategy content is essentially made up of two parts: strategic stance and strategic actions. Strategic stance is in the interpretation of Boyne and Walker (2004) the broad way in which an organization seeks to maintain or improve its performance, which proofs similarity to the fourth of Mintzberg's (1987a) definitions: 'strategy as a perspective'. This level of strategy is relatively enduring and unlikely to change substantially in the short term. The other component that makes up strategic content is more likely to change in the short term and identified as *strategic actions*. Strategic actions are the specific steps that an organization takes to operationalize its stance. One of the major advantages of dividing strategy content in strategic stance and strategic actions is that it provides a framework for analysis. Strategic actions can be assessed and comprehended in view of their explicitness and visibility. Strategic actions are the most visible projections of the strategy management of organizations, and hence are most prone to be subject in any research effort (Fox-Wolfgramm *et al.*, 1998). Strategic actions according to Boyne and Walker (2004) indicate how organizations actually behave, in contrast to strategies that are merely rhetorical, or intended but unrealised.

As Mintzberg (1987b) points out the difference between intended and realized strategies is important. Intentions that are fully realized can be called deliberate strategies. Those that are not realized at all can be called unrealised strategies. Emergent strategy is when a realized pattern was not expressly intended. Plans may go unrealised while patterns may appear without preconception. Plan is labelled as 'intended strategy' and pattern as 'realized strategy', as shown in Figure 6. 'Deliberate strategy', where intentions that existed previously were realized, is distinguished from 'emergent strategy', where patterns are developed in the absence of intentions. Realized strategies can be regarded as consistency of behaviour. Some state that strategy emerges within the general outline of a strategic plan, and from a foundation of activity taking place through the organization and according to a pattern of trial and error learning (Hatch, 1997). Others regard emergent strategies as strategies that come about without the explicit intention of managers but which result from the flow of more operational, day to day decision making (Johnson and Scholes, 1988). Various functions contribute in various unexpected ways to the emergence of strategies, ultimately affecting the realized strategy of the organization.

Figure 4 Deliberate and emergent strategies

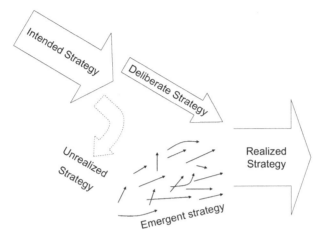

Source: Mintzberg (1987b).

Hence, combining the notions of De Wit and Meyer, Boyne and Walker, and Mintzberg three main dimensions are found relevant for the research:

1. Strategic context (what an organisation *can* do).
2. Strategic plans (what an organisation *wants* to do).
3. Strategic actions (what an organisation actually *does)*.

For each of these three dimensions a separate analysis is required, with the aim to understand whether WSS providers in different institutional contexts have different strategies.

4.7   Researching 'strategy' through typologies

Hambrick (1980) made an overview of the four main approaches to research strategies. He summarized the applicability, strengths and limitations of each approach, and its dependency on how strategy is defined and the aim of the research. The following four approaches are, according to him, available:

1. Textual descriptions.
Textual operationalisations of strategy actions are particularly useful in theory building. They are of limited use in theory testing, for primarily two reasons. First, they cannot be conducted economically in sufficient numbers to allow generalizability of results. Second, they do not allow reliable comparison across organizations or replication by other researchers. Rather they reflect the investigator's qualitative interpretation of each organization's strategy.

2. Measurement of parts of strategy.
Researchers rely in this case on one or a few key variables to portray strategic behaviour. The apparent limitations of such studies are that they do not capture the breadth of decision areas that constitute strategy. The greatest usefulness of such research may be that it allows a close, relatively precise, study of a limited array of strategic variables that, when better understood, can be more cogently incorporated into comprehensive theories and measures of strategy.

3. Multivariate measurement of strategy.
In this approach, researchers view strategy as a quantifiable interaction of a broad set of variables. Unlike the partial operationalisation of strategy discussed, the multivariate approaches take a comprehensive view of the construct. However, such an approach runs the risk of missing the concept of a central thread or internal logic underlying a strategy.

4. Typologies of strategies.
Strategic management theory provides conceptual typologies that are effective in identifying different types of strategies that are found within an industry. The typological approach is recognized as creating a better understanding of the strategic reality of an organization, since all types of strategy are viewed as having particular characteristics but a common strategic orientation. Research have found that companies within the same industry tend to use different types of strategies, but the different types of strategies can be classified into a limited number of categories. In fact, it is common practice in strategic management research to classify or group companies by strategic type and then identify strategy types within the given industry (Sriram and Anikeeff, 1991).

In view of the characteristics of the research at hand, the most appropriate approach to make strategies researchable is the latter one; the typological approach. The selection of the typological approach for this single industry research is supported by Ketchen *et al.* (1997). They statistically aggregated results from 40 empirical tests of the typologies-performance relationship. Their meta-analysis demonstrated that an organization's typology contributes more to performance explanation if these studies incorporate two criteria: the typology should be broadly defined, and the study should be executed in a single industry. The selection of the typological approach is an important step in the research design. However, as numerous typologies are available in strategic management literature it is prudent to select one typology. Several attempts have been made to empirically establish typologies of strategies (for example Ansoff and Stewart, 1967; Freeman, 1974; Miller and Friesen, 1977; Porter, 1980; Miles and Snow, 1978). Each strategic type is viewed as having its own distinct pattern of characteristics in these constructs. As two creators of a typology wrote (Miller and Friesen, 1977; page 264):

> The administrative situations that are described seem to form a number of gestalts. There is something systematic and ordered about the patterning of environmental, organizational, and strategy making behaviour attributes.

As the application of strategic management in the WSS sector is relatively new, the minimum requirement for a typology to be appropriate for the research is that it has proven extensively its value in other sectors. Hence, the typology needs to be applied many times by other researchers validating in this respect the value of the typology. Another minimum requirement is that the typologies should also enable to be related to different levels of performance as the thesis aims to connect strategies to performances.

In view of these minimum requirement two typologies are becoming particularly relevant, i.e. the typologies of Porter (1980) and of Miles and Snow (1978). Porter as well as Miles and Snow based their typology on comprehensive studies, with rich data and case studies. The typologies of Miles and Snow (1978) and Porter (1980) have been addressed by previous researchers as competing classifications of organizational strategy (Segev, 1987; Slater and Olson, 2001; Ruekert and Walker, 1987). Both the typologies of Porter and of Miles and Snow are also related to distinct levels of expertise and competencies, as well as to different levels of performance. However, while both classify strategies, the two typologies are different, each stressing somewhat different aspects of strategy. Hambrick (1983: page 698) suggested that:

> The two typologies are not incompatible; rather, their juxtaposition indicates the complex web of strategic options available to a business and the difficulty in trying to classify such options concisely.

Porter (1980) suggested that in any industry there are three potentially successful generic competitive strategies and one low profitability strategy:

1. Cost leadership, which aims to maintain lower costs than those of competitors, while maintaining quality of product. High volume sales are required for this strategy to succeed, as well as the associated economies of scale supporting the low cost focus.
2. Product differentiation, which aims to achieve industry-wide recognition of products and services which are different from those of direct competitors through quality or design or, perhaps, the type of customer service associated with them. Sometimes, customers are prepared to accept higher prices if there is perceived added value.
3. Specialization by focus, when organizations concentrate on a limited number of markets, products and geographical areas in which to compete. This could well approximate what is known as niche marketing.
4. The low profitability strategy is called the 'Stuck in the middle'-strategy. In this case the organization does not make a clear choice between any of the other three types of strategies, with the consequence that competitors continuously outperform the organization.

However, a major set-back of Porter's generic strategies for selecting it as the dominant typology in the thesis is that his typologies are primarily concerned with aspects of competition and maintaining a distinct competitive advantage. This focus does not align well with the WSS sector that has monopolistic features. Such is in contrast to the typology developed by Miles and Snow (1978).

The typologies of Miles and Snow have been applied by many researchers just like the generic strategies of Porter (1980), The diverse empirical studies that have applied Miles and Snow's model have contributed to identifying it as one having good codifications and prediction strengths. According to Miles and Snow (1978) their typology should be applicable in every industry as its Principal strength is the simultaneous consideration of the structure and processes necessary for the realization of a given type of strategy. This assertion has subsequently been validated by empirical research in multiple industries (e.g. Snow and Hrebiniak, 1980). Gimenez (2000) found that the Miles and Snow classification is a well-researched taxonomy and can be selected with less need to explore its operational status. He identified over 50 papers that have applied Miles and Snow's models in the period between 1987 and 1994. Croteau *et al.* (1999) basing themselves on the Social Sciences Quotation Index from 1989 to 1998 found more than 650 quotes in these years, illustrating the wide adoption of the typology. In a literature survey Croteau *et al.* (1999) claim in this respect that Miles and Snow's typology still has validity today in spite of its relative age. According to Desarbo *et al.* (2005) numerous authors have found the typology's longevity and excellence to its innate parsimony, industry-independent nature, and to its correspondence with the actual strategic postures of companies across multiple industries and countries. Desarbo *et al.* (2005) even state that the Miles and Snow framework continues to be the most enduring strategy classification scheme available.

Hence, the strategic typologies of Miles and Snow (1978) are selected to proxy the strategic actions of WSS providers in the thesis at hand. Though typologies of strategy are very convenient tools for studies, they may mask by their halo effect the distinctions between content and process. One approach to alleviate this effect is to decompose each strategic type into its basic facets, identify and amalgamate identical facets, recluster the facets into process and content groups and then operationalise and measure each facet independently. Using such approach might enable to focus on the typology from Miles and Snow (Segev, 1987).

4.8   The Miles and Snow typology

According to Miles and Snow, an organization shapes its strategy, structure and process (and their relative interrelationships) dynamically according to environmental changes and uncertainties, following what they defined an "*adaptive cycle*", which is composed by certain patterns of key decisions and solutions to three different problems: the *entrepreneurial problem*, the *engineering problem* and the *administrative problem*. The entrepreneurial problem deals with how the organization defines its product or service and target market. The engineering problem is offering an operational solution to delivering the services of the organization. While the administrative problem addresses the organization structures and processes to direct and monitor operations in order to reduce uncertainty faced by the entrepreneurial and the engineering problems.

According to Miles and Snow, organizations act differently when challenged with the adaptive cycle. In the Miles and Snow model, four different typologies may be distinguished according to the strategy pursued in their adaptation process: Prospector, Analyser, Defender, and Reactor. A description of the characteristics of the four different types of strategy is provided in Table 4.

Table 4 Strategic typologies adapted to the WSS sector

Orientation	Description according to Miles and Snow	Adaptation to the WSS sector
Prospector	A prospector organisation continually searches for new opportunities. It has a broad and flexible product/market domain and, hence, a broad technological base. It usually creates change and uncertainty in the environment. Its structure is characterised by a low degree of formalisation and routinisation, decentralisation, and lateral as well as vertical communication. It responds quickly to early signals of opportunities and is usually the first to enter a new product/market area. It is not necessarily successful in all of its endeavours, nor is it very efficient since product/market innovation is its major concern.	This WSS provider strives continuously to expand its market domain by targeting new customers and offering new services. It highly regards innovation and application of the latest technology. The internal organisation is flexible, decentralised and creativity is highly appreciated. Externally it is constantly on the search for new ideas and opportunities.
Defender	An organization with this orientation tends to have a narrow product/market domain. It will try to create and maintain a niche with a limited range of products or services. It also has a narrow technological base (because of its narrow domain). It does not attempt to search outside its domain for new opportunities. Hence, it becomes highly dependent on its narrow product/market area. As a result, it tries to protect its domain through lower prices, higher quality, superior delivery, and so forth. The structure of a defender is characterized by an elaborate formal hierarchy and high degree of centralization.	This WSS provider is fully occupied by serving its existing customers. It is good in what it does and it is proud of the outstanding quality of its service delivery. All revenues gained are invested in further improving the existing processes. The company concentrates on doing the best job possible in its' assigned service area. The rest of the world only matters if it inflicts with doing that job.
Analyser	An organization with this orientation has characteristics of both the defender and prospector. It tends to maintain a stable and limited domain, while at the same time cautiously moving into a new domain only after its viability has been proven by prospectors ('second-in'). Analysers are imitators in such a way that they take the promising ideas of prospectors and successfully market them. They seek flexibility as well as stability. They adopt structures that can accommodate both stable and changing domains.	This company thinks before it leaps. It carefully scrutinizes first all options, analysing the experiences of others, before selecting the right one. This strategy has proofed successful to grow gradually. In innovations, this company is not the first adapter, but follows if it has proven its' value.
Reactor	This firm does not have long-term goals or articulated strategies, and, hence, no consistent pattern of behaviour. The organization is passive in dealing with various issues. It does not attempt to maintain a defined product/market domain, nor does it try to capitalize on viable environmental opportunities.	This is a WSS provider that puts its priority in being able to quickly react on external developments. It needs to, as the influence of external actors, as regulators, on the company is large. If the regulator sneezes, the WSS provider gets a cold.

Source: Miles and Snow (1978) and questionnaire for the thesis at hand.

When Boyne and Walker (2004) applied the Miles and Snow typology specifically to public service organisations they managed to highlight some important critical points in the use of the typology: false conflicts between strategy typologies that are supposedly competing but are actually complementary, and simplistic and one-dimensional classification systems that seek to locate different organizations in mutually exclusive boxes. The following Figure indicates this uni-dimensional taxonomic approach they reject.

Figure 5 The uni-dimensional taxonomic approach

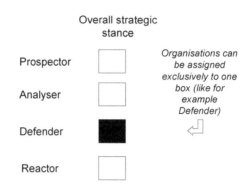

Source: Boyne and Walker (2004).

Boyne and Walker reject the uni-dimensional approach for two reasons. First, they find that an organization may have different stances in different spheres of its activities. In this respect Boyne and Walker defined 5 different types of strategic actions, and for each different strategic action the strategic stance might differ. Second, they share the notion that strategies can be mixed and combined. Strategies do not need to be mutually exclusive, so the attempt to satisfy this taxonomic criterion is inappropriate. Strategy variables, in their interpretation are continuous, not categorical, and a conceptual framework for identifying strategic archetypes should be consistent with this. Hence instead of locating for one strategic action precisely the strategic stance, Boyne and Walker propose to use scoring on a scale.

The second observation relates to the isolated character of the Reactor stance. The Reactor stance is that an organization does not have a consistent pattern of strategic actions, but is constantly changing its' course and action in reaction to outside pressures. The extent to which organisations have Reactor characteristics has by many researchers been linked to their performances. The more an organization has a Reactor stance, the less it performs. The isolated position of the Reactor stance has led that many researchers only included the other three archetypes into their research, i.e. the Prospector, Analyser and Defender stances. These three archetypes are by almost all researchers put on a continuum, in which the Prospector is the most progressive, while the Defender is the least progressive, and the Analyser is somewhat in the middle. As Ruekert and Walker (1987, page 17) argue, Analysers:

... are essentially an intermediate type between the Prospector strategy at one extreme and the Defender strategies at the other.

The other continuum is the extent to which a realized strategy had Reactor characteristics, which is separated from the Prospector-Defender continuum. The more a strategy has Reactor characteristics the more it can be labelled as inconsistent and constantly subject to interruption by outside forces. It is therefore, likely in the case of public agencies that are severely imposed by government and regulators that the organization would have its strategies imposed through the action of external agencies such as regulators. Based on the notion of the two continuums, a typological classification scheme is identified and depicted in Figure 6.

Figure 6 Typological scheme for the five defined strategic components

Strategic Archteypes	Scores on the continuums of:	Strategic Actions				
		Change markets	Change services	Seeking Revenues	Internal Organ.	External Organ.
	Def. – Prosp. (scale 1 to 10)	score	score	score	score	score
	Reactor (scale 1 to 10)	score	score	score	score	score

The higher the score in any of the boxes the more the strategic stance is classified as

•Prospector for the top row, or

• Reactor for the bottom row.

Source: Boyne and Walker (2004); modified by author.

A comparison of the strategies of operators with different attributes (like ownership, size or regulatory environment) can be presented through a scattered diagram (see Scatter Box 1). The vertical scale serves to place the WSS operator within the categories; Prospector, Defender or Analyser, while the horizontal scale serves to qualify the company as a Reactor or non-Reactor.

Scatter Box 1 The two continuums of the Miles and Snow typology

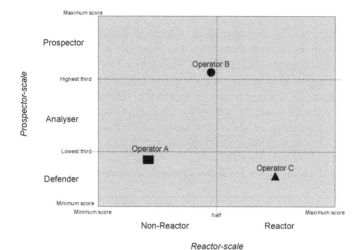

In the above-presented Scatter Box some positions of hypothetical WSS providers are included. In this example both operators A and C have quite strong Defender Characteristics, however, for operator C its Defender strategy is imposed on it as it has a high score on the Reactor stance, while for operator A it has selected this strategy out of its discretion. Operator B in this example has much more Prospector characteristics although its strategy is partly given in by external forces.

4.9   Strategy for public service organisations

Boyne and Walker (2004) have contributed further on the understanding of strategy by applying it specifically to public service organisations. In this respect, they identified three logical categories of strategies that are available to a public organization: change the environment (moving to a different market), change the relations with the existing environment (by altering services, revenues or external structure), or change itself (through modification of internal structure). Based on these categories, five strategic components emerge:

1.   Markets - Which clients does the company want to serve?
2.   Services - Which products/services does the company want to offer?
3.   Seeking Revenues – How does the company want to recover its costs through its   revenues?

The three just mentioned strategic components have a large similarity with the generic strategies of respectively focus, differentiation, and cost leadership, as defined by Porter (1980). Boyne and Walker propose to add two other strategic components next to the above three strategic actions. Their reason for inclusion of these two additional categories is because of the constraints that public organisations may face in altering the selection of client groups, services or prices. The strategic challenge many managers in the public realm have is to find better ways to deliver existing services in a fixed market with limited revenues. WSS providers must manage markets both upstream (donors and legitimacy) and downstream (clients and services). Strategy in only one direction (towards services and customers) does not fully capture the complexity of public management. Therefore two additional strategic components are especially relevant for the provision of public services, being:

4.   Internal organization - How to structure the internal organization?
5.   External organization - How to interact with external parties?

According to Boyne and Walker (2004), the combination of these five identified strategic components determines the strategy of public service organisations. The conceptual framework using the distinction in the five strategic components, developed by Boyne and Walker (2004) is selected to be applied in this thesis as it is particularly relevant for public service organisations, like WSS operators.

Strategic archetypes as developed by Miles and Snow are too complex to be measured effectively with only one single variable. A larger set of variables is necessary for appropriate reliability and validity assessment. As Jauch and Osborn (1981: 497) state:

> That is, using measurable variables to analyse strategy allows for integrating process and content, and conceptual and analytical issues.

In this respect numerous scholars have identified variables to assess strategies (like Conant *et al.*, 1990; Segev, 1987; Croteau *et al.*, 1999; DeSarbo *et al.*, 2005). More specifically, Schouten and Van Dijk (forthcoming) formulated variables for each of Boyne and Walker's strategic components for the WSS sector. In Annex 2 an overview is presented of the Miles and Snow typologies along the strategic components of Boyne and Walker.

In the following sections each of the five strategic components is further elaborated, including an identification of research variables.

4.9.1   Market strategies

Although WSS providers are restricted to certain geographical areas, they have some possibilities to extend their market. For example through bulk supply to neighbouring providers, a WSS provider is able to enter a new market subjected to different regulations. Also possibilities as inset appointments or common carriage are strategic actions for WSS providers to find new markets. Another possibility is to merge with or acquire other WSS providers, allowing a WSS provider also to serve users in other geographical areas. Conversely, in theory, a WSS provider can change its market by withdrawing from a particular geographical area within its service region.

Table 5 provides an overview on the variables related to the market strategies, defined by various researchers when assessing the Miles and Snow typologies. The column on the right of the Table presents the derived variables found relevant for the research.

Table 5 Variables relevant to market strategies

Scholars	Variables identified in other research for the strategic 'Market' action	Derived variables for the research
DeSarbo *et al.*, 2005	Knowledge of customers; Integration of marketing activities; Skills to segment and target markets; Effectiveness of advertising programs; Market sensing capabilities; Customer linking capabilities; Ability to retain customers; Marketing planning process	**1. Connections;** **2. Inset appointments;** **3. Bulk water;** **4. Mergers and acquisitions**
Conant *et al.*, 1990	Product-market domain (shared with the Product strategic action); Success posture; Surveillance; Growth	
Segev, 1987	Product-market breadth (shared with Product strategic action); Market share; Rate of growth	
Schouten and Van Dijk, forthcoming)	The sale of bulk water; Common carriage; Inset appointments; Unregulated supply; Mergers and acquisitions; International perspective	

4.9.2   Products/services strategies

According to the interpretation of Schouten and Van Dijk (forthcoming), the strategic products/services component concerns the delivery of other types of services and products to the client group that is already served by drinking water. WSS operators could diversify their products and services by offering different qualities of water or by offering also non-water related services. Such could include affiliated services as sewerage, but can extent to any service that the WSS provider might see as an interesting business opportunity or valuable endeavour.

Table 6 provides an overview on the variables related to the products and services strategies, defined by various researchers assessing the Miles and Snow typologies. The last column on the right of the Table presents the derived variables found relevant for the research.

Table 6 Variables relevant to products and services strategies

Scholars	Variables identified by other scholars on the strategic action of Products and Services	Derived variables for the research
DeSarbo *et al.*, 2005	New product development capabilities; IT systems for new product development; Technology development capabilities; Predicting technological changes	**1. Quality;** **2. Product portfolio**
Conant *et al.*, 1990	Technological breadth; Product-market domain (shared with the Market strategic action)	
Segev, 1987	Product-market breadth (shared with the Market strategic action); Product innovation; Quality	
Schouten and Van Dijk, forthcoming)	Core business of water; Other services than water	

4.9.3   Seeking revenues strategies

A major part of the strategy focus of any organization is ensuring that they have sufficient revenues to compensate the costs. Hence, changes in the efficiency in the use of resources and control of costs are categorized as part of this strategic action. The company could pursue to save on the costs or to increase the selling price. Potential strategic actions in terms of efficiency on the cost side are indicated with cost reduction programs, organizational restructuring, outsourcing, and technical innovations.

Table 7 provides an overview on the variables related to the seeking revenues strategies, defined by various researchers assessing the Miles and Snow typologies. The column on the right of the Table presents the derived variables found relevant for the research.

Table 7 Variables of 'seeking revenues' –strategies

Scholars	Variables identified by other scholars on strategic 'Seeking Revenues'-actions	Derived variables for the research
DeSarbo *et al.*, 2005	Effectiveness of pricing programs; Manufacturing processes; Cost control capabilities	1. Profit allocation;
Conant *et al.*, 1990	Technological goal	2. Efficiency;
Segev, 1987	Price level; Long range financial strength; Investment in production; Profitability; Operational efficiency	3. Asset management;
Schouten and Van Dijk, forthcoming)	Investment; Operational efficiency; Asset management; Gratuities policy	4. Gratuities

4.9.4   Internal organisation strategies

The strategic internal organisation actions refer to variables such as shaping the organizational structure, culture or style of leadership, the process of formulation and implementation of strategic planning, the use of pro-active strategy formulation instruments as scenario planning, and the adoption of performance measurement systems. Variables include defining missions, redesigning production processes, and using systems of operational management to rectify problems. For example, providers might choose to have a very flat organization with a lot of autonomy of the departments and employees.

Table 8 provides an overview on the variables related to the internal organization strategies, defined by various researchers assessing the Miles and Snow typologies. The column on the right of the Table presents the derived variables found relevant for the research.

Table 8 Variables of 'internal organisation'-strategies

Scholars	Variables identified by other scholars for strategic 'Internal Organisation'-actions	Derived variables for the research
DeSarbo *et al.*, 2005	Production facilities; IT systems for functional integration; IT systems of internal communication; IT systems for tech. knowledge creation; IT systems of internal communication; Integrated logistics systems; Financial management skills; HR management capabilities; Profitability and revenue forecasting	1. Innovation; 2. Knowledge and skills; 3. Training; 4. Strategy formulation; 5. Decentralization; 6. Empowerment; 7. Marketing
Conant *et al.*, 1990	Technological buffers; Dominant coalition; Planning; Structure; Control	
Segev, 1987	Technological progress; Control system level; Resources level; No. of technologies; Professionalization; Internal level analysis; Level of risk; Pro-active managerial style; Size of strategy making team; Centralization; Mechanism; Organizational size; Organizational age	
Schouten and Van Dijk, forthcoming)	Innovation; Management characteristics; Professionalization; Size of strategy team; Decentralization; Empowerment; Marketing policy	

4.9.5   *External organisation strategies*

External organization refers to the choices providers make in setting up inter-organizational relationships, like collaboration, networks, consortia and joint ventures, partnerships and outsourcing activities to private or non-profit parties. WSS providers are increasingly aware of civil society movements due to the perception of their potential impact on the business. A growing number of WSS providers engage in corporate citizenship programs with stakeholders to address social and environmental problems. Indeed, public (and even shareholder) expectations of corporations to deal with these problems in the communities where they operate have risen dramatically over the past decade, at the same time that the roles of national and local governments have been shrinking.

Table 9 provides an overview on the variables related to the external organization strategies, defined by various researchers assessing the Miles and Snow typologies. The column on the right of the Table presents the derived variables found relevant for the research.

Table 9 Variables of 'external organisation'-strategies

Scholars	Variables identified by other scholars for strategic 'External Organisation'-actions	Derived variables for the research
DeSarbo *et al.*, 2005	Durable relationship with suppliers; Channel bonding capabilities; Knowledge of Competitors	1. Environment; 2. Suppliers; 3. Benchmarking; 4. Regulator; 5. Partnerships.
Segev, 1987	Uncertainty; Dynamism; Complexity; Active marketing; External analysis and level	
Schouten and Van Dijk, forthcoming)	Image of the organization; Regulatory imposition; Outsourcing policy; Benchmarking; Inter-organizational partnerships	

4.10   The relation between 'strategy' and performance

One of the main reasons to select 'strategy' as the intermediate variable in the research design is the integrative nature of strategies and its relation to performance. This argument is reflected by Poister and Streib (1999: 310) who claim that strategies:

> embrace the entire set of managerial decisions and actions that determine the long-run performance of an organisation.

The main theory that identifies the link between strategies and performance is the Strategic Choice perspective that has spurred significant, systematic investigations of the influence of strategies on performance. Theory about strategies claims that organizations that exhibit certain types of strategy perform better than organizations with a different type of strategy[19]. Rumelt (1991) empirically validated this stance in his influential study over about 4,500 business units over a 4 year period. He found that the most important source affecting the performance of any business unit is the conduct of the business unit itself. The sector in which the business unit is operating is a much less important source and influence from the mother corporation is even quite important, according to Rumelt's calculations.

Almost two decades ago Zahra and Pearce (1990) did not find many references in their review of empirical enquiries on the relation between the Miles and Snow typologies and performance. Even more, Zahra and Pearce rejected any simple relationship on the evidence of these studies, and urged for further work to be conducted. Since the writing of Zahra and Pearce, numerous scholars have responded to this call and have reflected on the relation between the Miles and Snow strategic typologies and performance in many sectors. Doty *et al.* (1993) calculated that 24% of the overall variation in organizational performance can be predicted by using the Miles and Snow typology. In Annex 1 an overview is provided of a selected number of studies on the relation between strategic archetypes and performance. Unfortunately none of the studies executed has taken place in the WSS sector. Some of these studies will be highlighted in the following section, since they provide a means to formulate an appropriate hypothesis for the WSS sector.

The original contention of Miles and Snow (1978) was that Defenders, Analysers and Prospectors perform equally well and are superior to Reactors, due to the Reactor's lack of a stable strategy. Several empirical investigations have confirmed this hypothesis. Conant *et al.* (1990) found that in the health industry Defenders, Prospectors and Analysers had an equal performance in terms of profitability, but that all three types outperformed Reactors. Also the conclusion from Bahaee's (1992) analysis of 82 responses from the regional airline industry was similar. The same conclusions were found by Parnell and Wright (1993) for catalogue and mail-order houses based on the input of 104 respondents. Apart from concluding that Reactors were outperformed by the other three types, they also found Prospectors to be the best performing companies in terms of sales growth. Analysers, on the other hand, were found to be the best performing companies in terms of return on assets. A scholar that confirmed more recently the underperformance of Reactors is Giminez (2000). He

---

[19] Criticasters of strategic management literature argue that strategies in itself reduce the effectiveness of the organisation. According to them, an overemphasis of the value of strategies may stifle creativity in organisations, especially if strategies are rigidly enforced. In an uncertain and ambiguous world, fluidity can be more important than a strategic compass. Also when a strategy becomes internalised into a corporate culture, it can lead to group-thinking, and may cause an organization to define itself too narrowly (Miller and Cardinal, 1994).

showed that firms that adopted Defender, Analyser and Prospector strategies produced a better performance, especially in terms of turnover growth, than Reactor ones.

Although the conclusions of the above research support the original contention of Miles and Snow, other researches provide conflicting evidence of the relation between performance and strategy. Giminez (2000) stated that the relation between strategy and performance is less clear-cut than the outcome of his study would suggest. Variables both of external and internal nature add complexity to this relationship, such as size (Smith *et al.*, 1986), external environment (Snow and Hrebiniak, 1980; Hambrick; 1983; DeSarbo *et al.*, 2005), executive characteristics (Thomas *et al.*, 1991; Ramaswamy *et al.*, 1994), technological deployment (Croteau and Bergeron, 2001), knowledge orientation (Truch and Bridger, 2002), and the strategy formation process (Slater *et al.*, 2006). Below these studies are elaborated on.

Size. Smith *et al.* (1986) collected data from 47 electronic manufacturing firms. They assessed the performance (sales growth, profits, return on total assets, and overall performance) by asking the respondents to identify their company's relative performance compared to other firms in their region or industry. The data supported Miles and Snow's contention that Analysers, Prospectors and Defenders outperform Reactors. An interesting result was obtained on the relations between strategy, size and organization performance. Small Defenders outperformed Analysers and Prospector, Prospectors performed better as medium to larger firms, and Analysers performed better as large ones. This could indicate, according to the authors, that Miles and Snow might have captured different stages of strategy development rather than a typology of alternative strategic behaviours.

External environment. Whether the performances of the four typologies depend on the external environment has been researched by several authors. Snow and Hrebiniak (1980) tested the Miles and Snow typology in the automotive, plastics, air transportation, and semiconductor industries. As a surrogate for overall organizational performance, they used an objective measure of profitability: the ratio of total income to total assets. Their data, collected from 247 managers in 88 companies, supports the original contention of Miles and Snow Defenders, Analysers and Prospectors performed equally well and were superior to Reactors in three of four industries. In the fourth industry (air transportation) that was highly regulated, Reactors performed best.

Also Hambrick (1983) studied the relation between the typologies, performance and environment. In a study over 1,452 organizations over multiple industries, Hambrick found that defenders outperformed prospectors in stable mature and non-innovative industries, while prospectors performed better in innovative and dynamic environments. Prospectors presented higher product R&D expenses and marketing expenses as would be expected, while defenders produced high capital intensity, high employee productivity and low direct cost. Hambrick's study concentrated only on defenders and prospectors, failing to address the behaviour of the other two strategic types: Analysers and reactors.

DeSarbo *et al.* (2005) found that the original typologies did not sufficiently cover his survey population and made modifications to the 4 typologies related to countries were the organizations resided. They collected responses and performance indicators (profit margin, return on investment, market share, customer retention, sales growth and return on assets) from 549 organizations in China, Japan and the United States. They made their own composition into groups and found the performance of two groups superior, e.g. a group of defensive firms with marketing skills and a group of balanced prospecting firms. Two other groups were having the lowest performance, which were Asian-based prospecting firms with technology strengths, and US based firms with market linking and management strengths.

Executive characteristics. The relationship between executive characteristics (like age, education and tenure), strategic archetype, and performance, was the topic of study for both the researches of Thomas *et al.* (1991) and Ramaswamy *et al.* (1994). As indicators of performance they included sales, return on assets and return on equity, which were collected using secondary data. Ramaswamy *et al.* collected 109 responses in three US industries: computers and electronics, chemicals and petroleum. Focusing only on Defenders and Prospectors both studies found that a better co-alignment between executive characteristics and the strategic archetype of the organization resulted in higher performance.

Technological deployment. Croteau and Bergeron (2001) assessed the alignment between technological deployment, strategy and performance. Performance indicators (sales growth and market shares and financial liquidity position) were based on the respondent's perception. Based on data from 222 respondents employed by Canadian firms, they arrived at the conclusion that an outward technological profile contributes directly to organizational performance for the Analyser strategic archetype, while an inward profile of technological deployment contributes indirectly to organizational performance for the Prospector archetype.

Knowledge orientation. Truch and Bridger (2002) sought to see whether an alignment between knowledge orientation and strategic type results in better performance. They interpreted performance through three indicators: overall performance in the last year, return on investment in the last three years, and growth in volume of sales in the last three years. Based on 180 responses from a range of sectors including financial, services, professional services, telecom, education, IT, and the public sector, they concluded that up to a third of organizational performance may be impacted by correctly aligning knowledge orientation with strategic orientation, so the potential benefits of reviewing these areas and improving the alignment could result in significant performance improvements

Strategy formation process. Slater *et al.* (2006) endeavoured to establish a relation between the strategy formation process, the strategic typologies of Miles and Snow and performance. They took a sample from manufacturing and service businesses operating in 20 different industries. They excluded the Reactor stance in view of its low representation in the sample, and they changed slightly the other three archetypes into Prospectors, Analysers, low-cost Defenders and differentiated Defenders. Performance was assessed through asking the respondents about customer satisfaction, customer value, customer retention, sales growth, market share, profit. Based on 380 responses they found that Prospector performance benefited from a clearly articulated mission, while Analyser performance was harmed by it. Prospectors and Analysers both benefited from comprehensive alternative evaluation and none of the strategic types were harmed by it. Analysers were the only strategic type whose performance was enhanced by situation analysis. Prospector performance was harmed by a formal strategic formation process, while low-cost Defenders and differentiated Defenders' performance benefited from it.

4.11   Synthesis of the Chapter

The lessons learnt from the previous Chapters are taken into account in establishing the research design. From the previous Introductory Section it is clear that the dominant contemporary development in the WSS sector is the introduction of the neo-liberal agenda. The introductory section also indicates the lack of understanding amongst scholars of the value of these neo-liberal institutional changes. The thesis at hand aims to contribute to better understanding whether neo-liberal institutional changes indeed make a difference. In that sense, the thesis shares the same research objective as most of the existing research. However, the thesis adopts an alternative approach in trying to establish such. The main innovation is that it specifically incorporates 'conduct' into the research model. An important choice in the research design is to proxy conduct through the construct of 'strategies'. Strategy has many definitions, but the essence is that strategy "embraces the entire set of managerial decision and actions that determine the long-run performance of an organisation" (Poister and Straub, 1999: 310). Poister and Straub's definition clearly shows the integrative nature and the (causal) relation with performance, which validate the choice for strategies to be included in the research design. Moreover, the field of strategic management seems to increase its relevancy to the WSS sector; on the one hand because the WSS sector becomes more similar to other sectors, on the other hand because the strategic management field is extending beyond the purely competitive sectors.

To apply the construct of 'strategies' to a public sector, like the WSS sector, Boyne and Walker (2004) make an important contribution, which is selected to shape the conceptual framework of the thesis. Boyne and Walker subdivide strategy of public service organisations into five strategic components, e.g. market strategies, products/services strategies, seeking revenues strategies, internal organisation strategies, and external organisation strategies. According to Boyne and Walker, the combination of these five strategic components determines the strategies of public service organisations. Applying these five strategic components to the WSS sector, for each of them specific research variables can be identified that are to be integrated in the research methodology of the thesis.

Another important element shaping the research design of the thesis is to acknowledge that strategy is a multi-dimensional construct. Based on distinctions made by De Wit and Meyer (1994), Mintzberg (1987a) and Boyne and Walker (2004) three dimensions are found, each addressing strategy from another perspective. The three dimensions relate to the strategic context, which is in the thesis interpreted through the regulatory environment; the strategic plans, which in the thesis is interpreted through documented plans for long term operation of WSS provision; and strategic actions, which are assessed through the typological approach. Simply put, strategic context determine what WSS operators *can* do; strategic plans relate to what a WSS provider *wants to* do, and strategic actions to what a WSS provider actually *does*. With respect to these three dimensions, especially the relation between strategic actions and performance deserves specific attention as many scholars have established a relation between typologies of strategic actions and performance. However, for the

WSS sector such has not been subject to research. For the thesis this relation is however one of prime interest.

In sum, the innovation of the thesis lies in the inclusion of strategies as an intermediate variable between institutional (neo-liberal) changes and performance. The inclusion of strategies brings with itself several innovations to the existing body of literature. Only in few cases strategic management literature has been applied to the WSS sector, and insight in the relations between institutional changes and strategies on the one hand, and strategies and performance on the other hand will broaden the current understanding.

Section II. Research Design

Chapter 5 Analytical framework

This Chapter completes the research design section by using the insights gained in the previous chapter to operationalize the research.

5.1   Introduction

In this Chapter the Research Design section is completed. The Chapter starts with the identification of the Analytical framework. The framework is strongly based on the insights gained in the previous chapter. Based on the analytical framework, the research questions and hypotheses are formulated. Successively, the target population and the research techniques are defined. The Chapter ends with the identification of the limitations to the research.

5.2   Analytical framework

The previous chapter identified as the main research ambition to increase the understanding of the value of neo-liberal institutional changes in the WSS sector. Moreover, the chapter identified that by including 'strategy' as an intermediate variable in the research, this ambition could best be realized. 'Strategy' has accordingly three main dimensions, e.g. strategy context, strategic plan, and strategic action. Based on these notions, the analytical framework for this thesis is now constructed (see Figure below).

Figure 7 presents the main analytical framework of the thesis. The framework shows that the thesis explores the relation between neo-liberal institutional changes and performance through the intermediary of strategies. With some imagination one could visualize the incorporation of 'strategy' in the research design to bridge the gap between neo-liberal institutional changes and performance. The 'gap' (in knowledge) to be bridged refers to the existing ambiguous evidence on the relation between institutional changes and performance (see Chapter 3). The framework tests whether performance increases (or decreases) due to neo-liberal institutional changes might occur through the mediating effect of strategies. Compared to earlier research, the framework's main innovation is its integrative nature. Some of the relations between elements in the model have been tested separately but the integrative model itself has not been tested.

Figure 7 Main analytical framework of the thesis

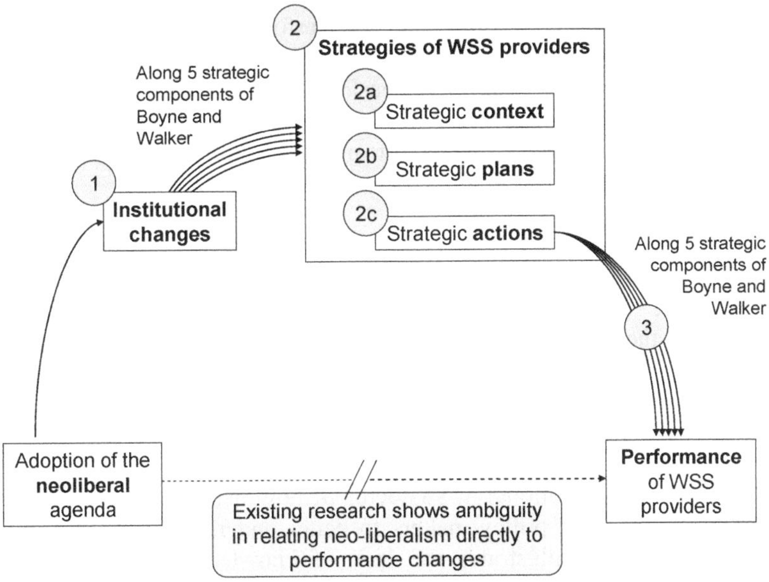

The following sections will further explain each relationship presented in Figure 7.

5.3   Research question and hypotheses

The research objective, as previously identified, leads to the development of one main research question and hypothesis to focus the research. The main question with its corresponding hypothesis is:

To what extent do neo-liberal institutional changes affect the strategies and performance of WSS providers?

It is hypothesised that indeed neo-liberal changes make a difference for both the strategies as the performance of WSS providers. An analysis of the relation between institutions, strategies and performance will support the decision making process on the value to pursue neo-liberal institutional changes in WSS provision. The main research question can be subdivided in 5 sub-questions following the path in the main analytical framework:

1. To what extent does the adoption of the neo-liberal agenda affect the institutions of the WSS sector?
2a. To what extent do neo-liberal institutional changes affect the strategic context of WSS providers?
2b. To what extent do neo-liberal institutional changes affect the strategic plans of WSS providers?
2c. To what extent do neo-liberal institutional changes affect the strategic actions of WSS providers?

3. To what extent do diverging strategic actions of WSS providers affect their performances?

For each of the five sub-questions it is hypothesised that indeed the changes make a difference to the strategic dimensions and the performance of WSS providers. As Prescott (1986) notes, a study like the one proposed in this thesis which analyses the relations between 'institutions – strategy – performance' will lead to satisfy a demand to understand better whether institutions are independently related to performance; whether institutions serve as a moderator of the relation between strategy and performance; or some combination of the two.

5.4   Target population

Although the thesis' primary focus is on WSS utilities from industrialized countries, the analysis and the findings may hold partial relevance for non-industrialized countries. Hence, some references to and cases from non-industrialized countries are included in the thesis. More specifically, the third Chapter included already a case study from Cochabamba, Bolivia, and in Chapter 8 a case study is conducted of the Caribbean island of St. Maarten, Netherlands Antilles. The 'Cochabamba' case is a landmark case in the international WSS sector of how neo-liberal changes can fail, spurring many scholars and policy makers to better understand the reasons for this failure. The case sharply illustrates the lack of knowledge of policy makers in engaging into neo-liberal changes. The St. Maarten case is included due to the unique opportunity of the researcher as external evaluator of the tendering procedure, giving him an opportunity to access inside knowledge and information crucial for the research, which would otherwise not be possible to attain.

For the purpose of the research, it is crucial to study WSS providers that are operating in different institutional contexts but which are relatively similar in their access to resources and environmental turbulence. A choice is made in the survey to restrict itself by only including the ownership element of the operator to delineate WSS operators subjected to neo-liberal pressures from other operators. Three distinct types of ownership arrangements for WSS providers are used in the thesis to distinguish and compare groups of WSS providers (see Figure 8). The three types of ownership can be placed on a continuum in which the three types can be regarded within the context of the thesis as an ordinal discrete scale of private sector involvement.

Figure 8 The types of ownership: a discrete ordinal variable

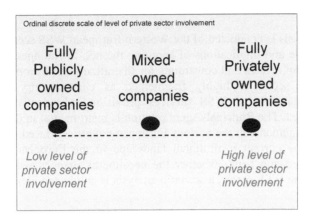

- *Fully publicly owned WSS operators.* The distinguishing feature this group has is that there is no sharing of ownership with non-public parties. The WSS operator can be an integrated part of a governmental organization, like a department or it can be loosely connected to the government through a publicly limited company. The operator can also be owned by different types of governments, for example it can be fully owned by a national governmental entity, but it can also be owned by groups of municipalities.
- *Mixed WSS operators.* The distinguishing feature of this group of operators is that there is a sharing in ownership by public parties and private parties. All options are included of a majority, a 50/50 distribution, or a minority of the shares in the hands of public and private parties.
- *Fully privately owned WSS operators.* This group only includes those operators in which the shares are 100% owned by private entities. This is again a heterogeneous group. Divestiture is included, in which there is a transfer of the ownership of the infrastructure, but also PSI type of arrangements are included, like concession contracts and lease contracts, in which essentially the management of the operations is 100% transferred (temporarily) to a private party. Especially, the WSS providers in England and Wales receive a lot of attention in the research in a comparison with other WSS providers (like from the Netherlands, Ireland, Scotland and Italy) due to their relative uniqueness as wholly privately owned.

5.5   Research techniques

An appropriate research technique is to be identified for each of the research sub-questions. Research techniques are the methods used for data generation and collection. Essentially research techniques are concerned with measurement, quantification and instrument building and with making sure the instruments are appropriate, valid and reliable (Oppenheim, 1992). The following sections will elaborate on the selection of the applicable research technique for each sub-question.

5.5.1   Neo-liberalism and institutional changes

The first sub-question refers to the influence neo-liberalism exercises over the institutions of the WSS sector. It is hypothesized that the adoption of the neo-liberal agenda affects the WSS sector's institutions. A full Chapter is allocated to this analysis (e.g. Chapter 6).

In effect, an analysis is conducted of the Western European WSS sector to understand better the rationale and implications of adopting the neo-liberal agenda. The analysis highlights the main drivers and constraints for liberalizing the European WSS sector. Since the WSS sector is highly fragmented as each country (or even each municipality) organises its' WSS services provision individually, a broad-brush approach is applied. The Principal-Agent relation is instrumental in distinguishing the institutional arrangements in the various European countries. Based on the Principal-Agent relation the current institutional landscape in the European WSS sector is mapped. To understand better whether the neo-liberal changes will also affect the future set-up of the WSS sector, a scenario analysis is conducted identifying plausible future end states.

5.5.2   Institutions and strategic context

The research question (2a) is to what extent do neo-liberal changes affect the strategic context of WSS providers? The corresponding hypothesis is that neo-liberal institutional changes indeed make a difference for the strategic context of WSS providers. Chapter 7 is dedicated to analysing the relation between institutions and strategies. This layer of analysis concerns an analysis of the neo-liberal implications on the strategic context. As Jauch and Osborn (1981: 491) state:

> Strategy can be defined as the combination (profile) of environmental, contextual, and structural elements affecting an organisation at any one time... At the most aggregate level, strategy is the profile of overall environmental, contextual, and structural complexity.

This definition allows for strategy being analysed at the aggregate level of the strategic context. Hence, this first level of analysis aims to define whether WSS providers operating in a different institutional context have a different strategic context. Strategic context is interpreted as managerial discretion, taking into account that WSS service providers are severely limited in their behavioural choices due to the regulatory and sector specific impositions. Hambrick and Finkelstein (1987) developed the concept of managerial discretion to refer to the latitude of managerial action. More specifically, managerial discretion pertains to the ability of a manager to exert his or her influence in strategies like resource allocation, product-market selection, or the launching of partnerships. The central idea is that individual managers have different ranges of decision making options available to them, determined by a combination of environmental, organisational and personal factors. To have a certain latitude of action is highly important for any manager of a WSS provider since without it an effective and efficient implementation would be considerably complicated. To address cases of individual users, a provider requires some room to interpret generally applicable rules, categories and laws. Teulings et al. (1997) identify three ways how government can give managerial discretion to providers:

1. Government can do it consciously by acknowledging that the execution of rules can only be done at lower levels due to the need to rely on the available professional and technological knowledge in the executing organisation. The law incorporates an autonomous character of the executing agency.

2. Managerial discretion can also be created due to a certain level of indefinites of the concept used by the legislator. For example, if the rules incorporate terms as 'fair' or 'affordable', the provider has the obligation to interpret that to its' local situation.

3. Thirdly, managerial discretion can be the result of conflicting legislation or vague and unclear governmental objectives. WSS providers are subject to sometimes-conflicting legislation from the supranational bodies, the national, provincial and municipal governments.

For the WSS sector the impositions of regulatory regimes on the managerial discretion are widely acknowledged. It is the task of the providers to stay within the managerial discretion to make the final decisions. The larger the managerial discretion, the more important is the role of the manager of the WSS provider for giving content to governmental policies. Managers of WSS providers cannot be expected to automatically serve the public interest properly and they need to be disciplined in such manner that the public interest is secured (Wetenschappelijke Raad voor het Regeringsbeleid, 2000). In practice WSS providers function in what can often best be described as a web of control and accountability (Schwartz, 2006). Each of the actors and groups affects the WSS provider through exercising different functions. Those functions are characterised by different relations and exert pressure on the companies. This can be in the form of non-political pressure, being impositions from clients, beneficiaries, and suppliers, who take action to demand higher standards of operational performance from a WSS provider (Israel, 1994). Or in the form of impositions from the political establishment or from regulatory or control agencies. Hambrick and Finkelstein (1987) illustrate the impact of environmental factors on the level of managerial discretion by comparing a manager of a public utility (like a WSS provider) with his or her counterpart at a medium sized microcomputer firm:

> The computer executive has legitimate options in the areas of pricing, promotion, production technology and locations, distribution, joint ventures and sales force incentives (to name a few); and realistically speaking, the utility executive's options in these areas are limited or even nil. Viewed another way, the two executives may be said to have the same possible domains of action, but greatly different ranges of discretion in many of those domains" (1987; page 372).

Interpreting the above quote, Hambrick and Finkelstein (1987) find that the leadership role of a manager of a WSS provider is restricted to that of a figurehead, due to the dominance of external environmental factors such as government regulation and environmental legislation. The restrictions made on managers in their latitude of action are often intentionally imposed to remedy the lack of a self-regulating market mechanism (Wills-Johnson et al., 2003; Robinson, 1997; Carney, 1990). As the provision of water remains crucial for a community in terms of health, economic development, externalities and environment; the service provider needs to oblige numerous laws, regulations and instructions that are put in place to secure the public interest. Moreover, characteristics of the business processes and the product of water leave little room for engaging into alternative actions. One could pose that due to the strict regulatory regimes and control, the strategies of the service providers are only dependent on the imposed regulation. Whether, the service provider has a private or a public owner would in this respect be of no concern. The regulatory regime is selected as an independent variable to represent the institutional context. It is noted that institutional context has a broader scope compared to the regulatory regime, as the institutional context incorporates all rules, both formal and informal, for WSS operators. The regulatory regime is more specific as it only includes the legislation passed by legislative bodies as well as the rules issued by administrative agencies (Reger *et al.,* 1992).

The effects of regulatory regimes on the managerial discretion have attracted scant empirical attention from researchers in strategic management or organisation theory. No empirical research was found from literature assessing or comparing the managerial discretion of WSS providers. However, an analysis of the regulatory constraints on the managerial discretion of WSS provider is an essential condition to study strategies WSS providers. Researchers have noted that different institutions may be conducive to certain strategic actions (Desarbo *et al.*, 2005). According to Reger *et al.* (1992: 189):

> Strategic management, organisation theory, and industrial organisation economics researchers have hypothesised that government regulation will affect strategic choice and performance. Despite the pervasiveness of regulation and the critical role of strategic choice in determining firm performance, the intersection of strategic choice and regulation has been largely ignored.

Within the context of the thesis, an analytical comparative case study is executed of the regulatory impositions and opportunities in a neo-liberal institutional context (i.e. England & Wales) and a traditionally public institutional context (i.e. the Netherlands).

5.5.3   *Institutions and strategic plans of WSS providers*
The research question relevant for this level for analysis is: to what extent do neo-liberal institutional changes affect the strategic plans of WSS providers? Hypothesis is that neo-liberal institutional changes indeed make a difference for the strategic plans of WSS providers. Chapter 8 is set-up to address the issue between institutions and strategic plans.

The next layer of analysis aims to identify whether the institutional context of WSS providers affects their strategic plans. Compared to the analysis of the strategic context, this level has a lower level of abstraction, by considering the goal preferences and predispositions of WSS providers. This level of analysis concerns the explicit strategic plans by WSS providers. The interpretation of strategy as a plan refers to the original contention how strategy was thought of. Many consider 'The Art of War' by the Chinese author Sun Tzu on military planning dating from the fourth century B.C. as the first time when the concept of strategy was systematically addressed (Mintzberg *et al.*, 1998). Even the work of 'strategy' itself is derived from the Greek verb 'stratego'; meaning to 'plan the destruction of one's enemies through effective use of resources' (Bracker, 1980).

Over time, strategies were not solely used any more for military organisations, but extended towards also to other organisations in society. In this interpretation strategies were thought of as "a plan designed to achieve a particular long-term aim" (Oxford Dictionary). These strategic plans are often written down explicitly in specific strategy documents like multi-year planning documents, mission and vision statements and company objectives. The distinguishing characteristic of these documents is that they are relevant for a relatively long period into the future, and encompass all of the organisation's operations. A specific case in which strategic plans are described is in bidding documents to operate and invest for a long period of time WSS services provision. These documents meet the criteria of being relevant into the far future and addressing all of the functionalities of the operations. There are several arguments that make the analysis of bidding documents particularly suitable for research of strategies in the WSS sector. Apart from the fact that there are many long term tendering events organised in the WSS sector for concession-like arrangements, the use permits a comparison of the plans of different organisations for the same situation. Any periodic or situational dependency is eliminated in the comparison.

Again an analytical case study is provided, although for this level of analysis it concerns a single case study. According to Yin (2003) single case studies can be very relevant if they have a certain unique character. The single case study originates from the Caribbean Island of St. Maarten. Although St. Maarten is outside the geographical primary focus of the thesis, it is included in view involvement of the researcher in the case. The researcher was directly involved in the case as he acted as an external evaluator of the bidding documents on behalf of the Island Government. This involvement allowed him to access data and gain unique information on the case, which would have otherwise been extremely difficult to collect.

5.5.4   *Institutions and strategic actions of WSS providers*

The relevant research question for this level of analysis is: to what extent do neo-liberal institutional changes affect the actually realized strategic actions of WSS providers? There is some evidence that the selection of strategic actions does not follow the highly rational pattern of strategic plans (Jauch and Osborn, 1981). As Poister and Streib (1999: 311) point out:

> As effective public managers know, organisations move into the future by decisions and actions, not by plans. If plans are not implemented in a very purposeful way, then the strategies will not take hold, no matter how compelling or inspiring the planning process.

Hypothesis is that neo-liberal changes indeed make a difference for the strategic actions of WSS providers. Chapter 9 is dedicated to the analysis between institutional changes and strategic actions.

As discussed in the previous chapter, strategic actions are estimated in the thesis at hand through the strategic typologies as developed by Miles and Snow. Still it needs to be identified which research technique is to be used to collect the data in this regard. Investigators followed different research techniques identify organisations along the Miles and Snow typology. Snow and Hambrick (1980) distinguish between four broad approaches for assessing the strategic archetypes of Miles and Snow:

1. Through analysis of secondary data of organizations (see for example Hambrick, 1983). Using this type of research, objective indicators are collected from an organization, as for example the percentage of sales derived from new products indicating the degree of innovation. In few cases researchers have opted to use secondary data related to objective indicators to assess the strategic archetype of an organization. A large setback of the use of objective indicators is that not all of the four archetypes of Miles and Snow will be easily found. In particular the

Reactor stance and the Analyser stance will be troublesome to identify as this research method in essence uses a one-dimensional conceptualisation of a multi dimensional construct.

2. Through interviews (see for example Ruekert and Walker, 1987). The use of interviews allows all four types of the strategic archetypes to be captured and it is somewhat objective. However, just like the external assessment method, interviewing is time consuming. Another disadvantage is that is only useful in case the sample is relatively small.

3. Through an expert panel (see for example Meyer, 1982). External assessment of strategic typologies involves an expert panel that conducts the strategic typing of a sample of organisations. Benefits related to the use of an expert panel are the impartial assessment, the capability to capture all four of the types and its' potential to be used for large samples. However, the use of expert panels is relatively time consuming. It can take a fair amount of time for experts to be identified, their involvement secured and a process developed by which classification decision can be made.

4. Through a survey (see for example Snow and Hrebiniak, 1980; McDaniel and Kolari, 1987; Segev, 1987; Zahra, 1987; Tavakolian, 1989; Shortell and Zajac, 1990; Karimi *et al.*, 1996).

Many researchers have attempted to map the strategic archetypes of Miles and Snow through a survey. This is also called 'self typing' as respondents are asked in a questionnaire to self-type the strategic archetype of their organisation. Two forms of self-typing exist: the "paragraph description" and multi-scaling. In a paragraph description, a respondent is asked to classify his organisation based on four paragraph descriptions, each reflecting one of the strategic archetypes. The paragraph description is often used as it allows for a relatively easy job for respondents to complete a questionnaire. Moreover, interpretation of the data is straightforward and it allows all four types of the Miles and Snow typology to be captured. Hence, especially when the research sample is relatively large, a paragraph definition is useful. Application of the paragraph description incorporates some disadvantages. One of these disadvantages is its' multi-exclusivity. It could very well be that respondents think their organisation has characteristics from several of the strategic stances but they are asked to tick only one the boxes. Another limitation is that a paragraph description is able to capture only few of the strategic dimensions typical of the adaptive cycle. One paragraph describing a strategic archetype over-simplifies the eleven adaptive cycle dimensions on which Miles and Snow have constructed the archetype. Typically, only two or three of the strategic dimensions explicated in Miles and Snow's adaptive cycle model are considered and evaluated in this approach (Conant *et al.*, 1990). To overcome the defects of the paragraph description, some researchers have added an item-based approach to the paragraph description (Conant *et al.*, 1990; Segev, 1987; Namiki, 1989; Smith *et al.*, 1986, Thomas *et al.*, 1991). In this addition a multi-item (Likert type), close-ended scale is developed which reflects the overall degrees to which an organisation's strategy conforms to the Defender, Prospector, Analyser, and Reactor archetypes. Cluster analysis is used to classify companies into any of the four strategic archetypes. Just like the paragraph definition, the item based approach is particularly useful in case of large samples. The multi-item scale allows all four archetypes to be captured in the research. But although the multi-item scale allows for many of the eleven adaptive cycle dimensions to be included, still a degree of simplification of the archetype constructs can occur. Also problems related to scale inconsistencies may occur as for example the number of items may vary by strategic archetype (nine for Defenders, eight for Prospectors, seven for Analysers and four for Reactors).

Examining the different approaches towards classifying the strategic archetype of an organization, the following observations can be made:

- Surveys have been the most widely employed method to research the Miles and Snow's strategic typology.
- Some studies have relied on single-item scales when operationalising what is recognized to be a multi-dimensional construct. Due to this perceived weaknesses later studies have relied on identifying Miles and Snow's archetypes using multi-item scales.
- Some studies do not capture all of the four strategic types. Interestingly enough, many studies that have operationalised the strategic stance of organizations, exclude one of the four strategic stances identified by Miles and Snow, e.g. the Reactor stance. The other three are in these studies set out over a continuum from Prospector-to-Analyser-to-Defender, while the Reactor is completely out of the focus of the researchers (see for example Thomas *et al.*, 1991).
- Especially interviews and expert panels are restricted to small sample sizes.

In view of these requirements a choice is made to use a survey as research instrument. In the survey a combination is made between the paragraph description and a multi-item scale assessment. This means that the respondent is not only asked to broadly identify which of the four typologies fits best his WSS operator (paragraph description), but also he/she is asked to respond to a list of items assessing elements of the typological description (multi-item).

5.5.5   Institutions, strategies and performance

The sub-research question that guides the last layer of analysis is whether different strategic actions consequence different performances. As theory from other sectors has indicated that the existence of a relation between performance and strategies is common, it is hypothesized that this is also the case in the WSS sector. Hence, the hypotheses are supporting the notion that neo-liberal institutional changes are indeed making a difference, at least to the extent that they will change the strategies of companies.

At this level the results from the mapped strategies (from the analysis of relation (2c) are compared to the performance. The aim is to identify whether strategies have an impact on performance. Problem is that there are severe limitations in the availability of performance data related to neo-liberal institutional changes in the WSS sector. Contract documents are frequently classified, and post-award negotiations between public and private parties take place behind closed doors. A time series data analysis on utility performance before and after privatisation is available for a handful of cases only, despite the notion that institutional changes are a change over time and should be assessed as such.

Most empirical enquiries have, hence, relied on benchmarking data to interpret the performance of WSS providers (Braadbaart, 2005). Crain and Zardkoohi (1978) view the availability of reliable performance data of WSS operators due to benchmarking schemes as beneficial to any research activity in the WSS sector. Studies from the USA use the same basic data set from periodical surveys of the American Water Works Association (AWWA). Studies from Asia are based on survey data from the Asian Development Bank (ADB). Studies in Africa are mostly made use of the data from the Water Utility Partnership's (WUP) SPBNET database. The International Benchmarking Review (WRc, 2001) identified some 160 benchmarking schemes covering at least 700 water/wastewater utilities in 110 countries. Annex 3 provides a short overview of a selection of benchmarking schemes in the WSS sector. Annex 4 provides an overview standard performance indicators used in benchmarking WSS providers.

In this research, the performances of the WSS service providers are assessed through existing benchmarking data from the English regulator OfWat and the Dutch water companies association VEWIN. Admittedly, to use these two data sets is an arbitrary choice based on the frequent use of benchmarking indicators in the WSS sector, and data availability. The OfWat benchmarking indicators comprise five key categories related to the drinking water provision, which are consolidated in one overall indicator called OPA (Overall Performance Assessment). Next, the data from OfWat comprise financial performance indicators which complement the service level indicators. In this respect, a comparison is made in the research of the Return on Capital Employed with the strategy scores. The performance indicators of the Dutch association VEWIN are differently arranged. The indicators used in the Netherlands are divided over 4 main areas: water quality, service, environment, and finance and efficiency. There is available data of in total 9 performance indicators from 10 countries from 2003 and from 2006. The Dutch have not calculated one consolidated performance index, but for the research at hand an attempt is made to construct a scale from a combination of performance indicators (see Figure below):

Figure 9 Performance, an ordinal continuous scale

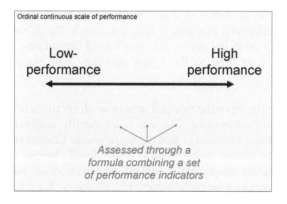

Chapter 10 is set up to identify the relation between strategic actions and performance of WSS providers.

5.6   Research limitations

Unavoidably in the research design numerous choices are made for the study at hand. On the one hand these choices give the study its' relevance and innovative character, on the other hand each of these research choices brings limitations in the use and interpretation of the study outcomes. Acknowledging the complexity of the problem to understand the effects of institutional changes and the determinants of performance, modesty should be taken in interpreting the results of the study. The research may be helpful to contribute additional understanding, but remains limited in generating any comprehensive conclusions. As White and Hamermesh (1981: 213) observe:

> A satisfactory answer to the question "What determines the level of a business's performance?" has proven illusive. Partial theories and answers abound, but none has offered a comprehensive explanation. Given the complexity of the problem, it is not surprising that there exist a number of largely independent areas of research that have attempted to explain performance. They emphasize different and usually singular explanatory factors, have different conceptual schemes, use different language and examine different organisation units.

The study at hand is in this regard no exception. Several limitations of the study are in the below sections identified which should be kept in mind when interpreting the study outcomes. In this section, for the most important choices made, these limitations are addressed. The following choices feature in this analysis of research limitations:

1. The choice of a cross sectional approach instead of a longitudinal approach.
2. The choice of 'strategy' to proxy the conduct of WSS providers.
3. The choice to use 'ownership' of the WSS provider as the determinant variable to delineate the various manifestations of neo-liberalism in the WSS sector.
4. The choice to focus on the transaction between the responsible and the management entity in analysing institutional changes.
5. The choice to use the analysis of variance (ANOVA) as the data processing methodology.
6. The choice to proxy performance of WSS providers through benchmarking indicators.

The limitations with respect to each of these choices are presented in the following sections:

5.6.1   The choice of a cross sectional approach

Within the research design the notion of institutional changes is central. The main question is whether institutional neo-liberal changes matter. This will essentially be investigated by comparing operators active in one (neo-liberal) institutional context with operators in a traditional (not neo-liberal) institutional context. However, institutional changes are a dynamic process happening over a period of time. A cross sectional study, like the one proposed in this thesis, depends on a static comparison, which can be typically addressed by comparing public with private WSS operators in a given period. Such a static comparison will not capture the full dynamics of the institutional change. Institutional changes are by definition a change over time, and needs to be addressed dynamically by looking at a given operators' evolution and change between its private and public stages. A weakness of the longitudinal studies is that they may fail to control for period specific effects.

Also, the outcome of the cross sectional comparison may suffer since other factors apart from the institutional context affected by the neo-liberal agenda may also change. It is a problem for researchers to isolate the neo-liberal institutional change from other changes. Several factors, independent of the private-public distinction, intervene in the relationship between privatisation on efficiency (Villalonga, 2000). Moreover, some of these factors do so in a dynamic way, thus affecting the timing of privatisation effects. In many cases neo-liberal change within the WSS sector does not only entail a shift in ownership, but also a shift in subsidy structures and financing mechanisms. For example, when the English and Welsh WSS operators were privatised they received a substantial grant for a smooth change. Such change in financing mechanisms might influence also the strategies and the levels of performance of the operators. Hence, conclusions that a different ownership type has indeed an impact on the performance should be interpreted with caution.

5.6.2   The choice for 'strategy'

A conscious choice has been made in the thesis to proxy the conduct of WSS providers through the construct of 'strategy'. However it is recognized that this choice brings several limitations to the research and that some people have developed degrees of sceptism with respect to the use of strategies (De Wit and Meyer, 1994).

Criticism relates to numerous aspects on the use of the strategies in research. An important criticism is that despite the vast amount of strategic management literature there is no trade off between rigor and relevance in strategic management research (Freeman and Lorange, 1985). Strategic management theories tend either to be too narrow in focus, or too general and abstract to be applicable to specific situations. The argument is that strategic management theories arrive too late and are only able to solve yesterday's problems. This explains according to the criticasters the relatively short lifespan of many theories of strategic management. The above listed criticism necessitates also caution in the use in the thesis at hand of notions from strategic management literature, especially since the application of strategic management literature to the WSS sector is new. Within the context of the thesis, specifically developed frameworks from Boyne and Walker, Miles and Snow, and De Wit and Meyer operationalize the study at hand.

Other criticism relates to the influence strategic management literature may have on the day-to-day work of managers. The argument is that strategies do not have relevancy in practice and are only a paper tiger. In this respect Mintzberg (1973) found that senior managers typically strategize in an ad hoc, flexible, dynamic, and implicit way. Hence, managers are in their day-to-day work not concerned with the theoretical sense of direction written down in their strategy documents.

5.6.3   The choice for 'ownership'

One important element in the study at hand is that a comparison is made between a group of WSS providers subjected to neo-liberal institutional changes, and group of providers isolated from these changes. The thesis uses as the determinant variable to delineate the two groups one variable being the 'ownership' of the WSS provider.

A problem for the research is that it has to cope with the different manifestations of neo-liberalism. This problem specifically occurs in distinguishing the groups for comparison. Theoretical distinctions are satisfyingly crisp; empirical comparisons are messy, tentative, and hedged about with conditions. In previous research the group subject to neo-liberal institutional changes is often interpreted as "private companies", while the group isolated from these changes is interpreted as "public companies". However, this does not entirely give a clear-cut distinction. 'Private' normally refers to formal and profit making enterprises, but can also denote any organization that is not public (Budds and McGranahan, 2003). There are the large, commercial and multinational WSS providers like Suez and Veolia, alongside small scale and/or informal operators. Another type of a 'private' organization, or at least of a non-public organization, is the non-governmental organization (NGO) that is sometimes involved in WSS services delivery. The distinction between public and private becomes even more blurred in case of autonomous public limited companies that act as regular private companies, while their shares are owned by the government. A final example of an arrangement which is difficult to categorize as either 'public' or 'private' are the mixed or jointly owned utilities in which both public and private parties hold a share.

Admittedly the classification of groups of WSS providers in the research design based on only one criterion (ownership) is a severe simplification. Chapter 3 introducing neo-liberalism clearly identified that a shift in ownership is only one part of the adoption of the neo-liberal agenda. In this respect, the thesis cannot claim to state any general conclusion on the relevance of neo-liberalism to the WSS sector, as it is only addressing one manifestation of it (shift in ownership of WSS providers).

5.6.4   The choice for the data processing methodology

It is recognized that care should be taken in selecting the right statistical method to analyse the data. Different methods can be selected, and a choice is made in this thesis to use the analysis of variance (ANOVA) as the most suitable. Two recent studies illustrate in this regard the critical importance to select the right statistical methodology. These studies are the two consecutive studies in 1999 and 2002 conducted by Estache and Rossi, and the Kirkpatrick, Parker and Zhang study, dating from 2004.

Estache and Rossi (1999 and 2002) assessed in two separately published studies whether private operators outperformed public WSS operators in the Asia Pacific region. In their first study, they applied a stochastic cost frontier analysis and found that private operators were consistently more efficient than public ones. However, three years after their first study they changed the research methodology of the study by including more variables in their data analysis. By applying error components and technical efficiency models in the stochastic cost frontier analysis on the same data set as three years earlier, their conclusions also changed. This time they concluded that efficiency was not significantly different between private and public operators.

In the research from Kirkpatrick et al. (2004) a series of data analysis methods were used, each indicating a different outcome. The first analysis they executed was by calculating the average (and standard deviation) of a range of performance measures for both a sample of private as a sample of public WSS providers. They concluded based on this analysis that although the service levels were similar, the efficiency of private WSS providers was superior to public WSS providers. Private WSS providers had a higher labour productivity, a lower Unaccounted for Water, a lower proportional spend on labour in operating costs; they were more economic in its use of fuel and chemicals, and they achieved a slightly higher capital utilisation. However, the private operators charged 82% higher compared to public operators. Kirkpatrick,

Parker and Zhang found that interpretation of the just presented conclusions should be conducted with care, as the standard deviation confirm a high degree of variance in performance within both the public as the private operators. The differences between private and public operators may be due to the different way of operating of private actors compared to public ones, but may also be caused by the possibility that certain geographical areas might be more eligible for private sector involvement, or by the larger scale of the private operators. Hence, to make the conclusion more robust, an additional analysis was performed by generating data envelopment analysis (DEA) efficiency scores. For this analysis, they selected from their initial sample 71 operators of which 7 were privately owned. Additional data were included that might affect the performance indicators, including GDP per capacity, country-level water resource availability and a 'freedom index'[20]. Result from this second step was that all private WSS operators scored higher than 80% relative efficiency while 5 out of the 63 public operators scored lower than 80%. Hence, similar to the first step of analysis using averages, also the DEA analysis suggested that private WSS are more efficient compared to public operators. An additional insight from the DEA analysis was that many African public WSS operators seem to perform relatively efficiently. Kirkpatrick, Parker and Zhang conducted also a third analytical step. Lastly, they made a stochastic cost frontier analysis on the data set. This time the conclusion was contrary to the results from the previous analyses. It showed that private operators had higher costs to produce the same output, although the differences between private and public operators were not statistically significant.

The studies from Estache and Rossi, and from Kirkpatrick *et al.,* show that the statistical data analysis method selected has apparently a large effect on the outcome of analysis. Scholars should be very cautious in selecting their data analysis method. Within the research methodology presented, especially for relation 3a, analysis of variance (ANOVA) is selected as the data processing methodology. For the study at hand, there is no a priori reason that the ANOVA is not the most suitable data processing method. However, it is still valuable to point out that the outcomes of the study at hand should not be taken for granted without a thorough understanding of the data analysis methods used.

5.6.5   The choice for benchmarking indicators
The study at hand ultimately intends to understand the value of neo-liberal institutional changes for the WSS sector. In this respect, an analysis of performance changes is crucially important and conducted in the last phase of the research. A limitation to the research is that performance of WSS providers is not easily captured. Within the context of the research a choice was made to proxy the performance of WSS providers through benchmarking indicators. To some extent this addresses the nature of performance of a WSS provider as multi-dimensional and highly locally dependent. However, in interpreting outcomes of the analysis caution is needed in view of the complexity of the construct of performance in a public sector industry like the WSS sector.

Compared to profit oriented industries, the issue how to assess performance is more complex in a public sector like the WSS sector. WSS providers are markedly different

---

[20] The freedom index is a tool developed by the Fraser Institute that takes account of policies within countries affecting the size of government, legal structure and security of property rights, access to sound money, freedom to trade, and regulation of credit, labour and businesses. The freedom variable was included to capture wider governance or regulatory effects on performance of water operators, which might otherwise have been attributed to ownership.

from competitive industries as they are in principle non competitive and in many cases not profit oriented. As stated by Ring and Perry (1985: 282):

> Thus performance criteria for public organizations, the ultimate basis for judgments about management's performance, also appear to differ qualitatively from those of the private sector.

Speckbacher (2003) noted that profit oriented organisations benefit from three important features that reduce the complexity of performance measurement compared to non-profit organisations: (1) the owner as the privileged interest group, (2) with homogeneous needs, (3) which are relatively well expressed financially. Neither of these three applies to the WSS sector. According to Brown and Iverson (2004) the connection between performance and financial well-being does not necessarily apply to WSS providers. The performance of a WSS provider is determined by meeting the expectations of the community stakeholders, while profit organizations can focus on satisfying the demands of the owners of a firm. Also a National Research Council study (NRC, 1995: 36) shares this notion that any assessment of performance of an infrastructure industry like the WSS sector depends on the stakeholders:

> ... performance was the degree to which infrastructure provides the services that the community expects of that infrastructure [and] can be defined as a function of effectiveness, reliability and cost...

Due to the dependency of stakeholders to interpret performance, the selection of appropriate performance indicators is very locally dependent, serving a multitude of constituencies whose goals and needs may be quite heterogeneous. An additional complication is that many demands from stakeholders are difficult to measure, such as environmental sustainability and social responsibility (Robinson, 1997; Ogden, 1995). Van Dijk and Schwartz (2005) see the question, which factors measure performance as one of the most crucial and relevant in the current practice of benchmarking in the WSS sector. As Andrews *et al.* (2006) note there is a lack of attention by public administration academics to the issue of performance, although in recent years a literature on this topic is slowly emerging (Boyne, 2003). Alegre (2000) points out that in the early 1990s, when the International Water Supply Association (IWSA) selected the topic "Performance Indicators" for one of its world congresses, no abstracts were submitted and the topic was cancelled from the list of presentations; an indication of how little or no interest the topic raised in the sector at that time. Three of four years later, an inquiry by IWSA over 150 senior members in WSS providers worldwide indicated that performance indicators was by far the main topic of greatest interest in the sector then.

In case researchers choose to use more than one indicator, they face the problem to establish an ordinal continuous index scale based on a basket with abstraction rates, tariffs, labour efficiency, and numerous other indicators. However, it is not easy to specify the relevant models, especially in an industry where operating conditions differ so much from company to company. WSS providers vary along dimensions aside from organizational form. They differ by factors such as size and dispersion of the population served; in the scale and age of their capital equipment; in costs paid for labour, machinery, water, energy, and finance; in the quality of available water supplies; and in how much they treat the water before pumping it to customers. Since some of these features might differ systematically between public and private WSS providers, it would likely be misleading simply to divide the WSS suppliers into "public" and "private", to find the two average costs, and to attribute whatever difference there is to ownership effects (Donahue, 1989). Consequently spurious results will appear. Moreover, the results will often appear inconclusive once the standard econometric testes are applied. The British regulator for instance uses 'judgment' in its' models, as do virtually all users of econometric models, but the application of judgment is not straightforward when the value of the underlying models is unclear. Only in situations where there is indeed a comparable local situation a comparison between WSS providers is valuable, but unfortunately these situations seldom arise. Even within a country it is extremely difficult to standardize sufficiently to make useful cross-company comparisons (Robinson, 1997; Ogden, 1995).

5.6.6   The choice for the market to fulfil the governmental mandate

Within the thesis, a choice is made to focus on the relationship between the responsible entity (the government) and the management entity (the operator). However, the choice to select the relation between the responsible entity and the management entity implicates that no generalisation can be made towards the entire WSS sector. The WSS sector consists of more than one market and it is important to not simply view the WSS sector as one amorphous whole but as a set of interrelated markets that require detailed consideration on a case-by-case basis. In segmenting the markets of the WSS sector, a rough division can be made in four different types of markets, including the market to fulfil the governmental mandate that is the focus of the thesis. The four markets and their relation to the WSS provider are depicted in Figure 10.

Figure 10 Conceptual regulated transaction framework of the WSS sector

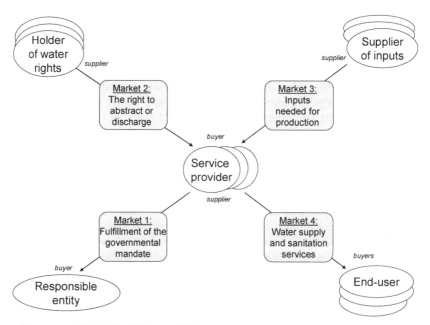

Source: Schouten and Van Dijk (2008); modified by author.

The consequences of an institutional change may differ from one market to another. The net effect of the institutional change for the entire sector may be opposite to the initial objective of the reform measure. To clarify this issue some attention is given in the following paragraphs on the existence within the WSS sector of separate markets.

In the following sections each of the remaining three markets is addressed. For each of the identified markets three main questions are to be addressed, just like for the previously described market to fulfil the governmental mandate; to what extent is privatisation relevant? To what extent is liberalization relevant? And to what extent is PSI relevant?

Typically the entity that has the mandate to provide WSS service provision to a defined region is the municipal council, although there are some notable exceptions - especially where WSS services are provided at a regional scale and/or by service providers that operate under private law. To fulfil its mandate the municipal council has several opportunities, including privatisation, liberalisation and PSI (see Table 10).

Table 10 Neo-liberal manifestations for the market to fulfil the mandate of the government

Arrangements relevant for the market to fulfil the governmental mandate	Is it liberalization?	Is it PSI?	Is it privatisation?
Divestiture	Possibly	No	Definitely
Benchmarking	Definitely	No	No
Delegation contracts	Possibly	Definitely	No

In case of privatisation, the municipal council decides that the ownership of the infrastructure and its management is transferred from public into private hands (divestiture). The newly private firm in this respect has the obligation to fulfil the mandate of the government, e.g. to serve a dedicated region. In case the government decides to organise competition between several potentially interested suppliers, privatisation become mixed with an element of liberalisation. However, the degree of competition is relatively low in this respect as it is only appearing at one moment. Once the private operator has acquired the mandate to serve the region, it can act as a (private) monopolist.

Most often when liberalisation in general is discussed, the discussion refers to delegation contracts concluded between the municipal council and private parties. The delegation contracts are a type of competition *for* the market, and in this case liberalisation goes hand in hand with PSI. The municipal council can decide to ask a private party to act on its' behalf as the provider of the full range of services within their mandate, or it can break up its' mandate in smaller parts by isolating some parts of the vertical chain of activities in the water treatment distribution process. Föllmi and Meister (2002) argue that delegation contracts have several drawbacks, particularly in the capital-intensive WSS industry. Since parameters as technology, water quality, demand or legal aspects alter over time; the setting of prices in advance is dominated by uncertainty. To handle this problem, the public authorities could regulate the industry or repeat the auction frequently. However, Armstrong *et al.* (1994) indicate that, besides the considerable costs of frequent auctioning, also shorter holding periods of the monopoly right undermine the incentives to invest into the network infrastructure.

Another option available to liberalise this transaction, without engaging into PSI, is to establish comparative competition. Comparative (or quasi) competition has come to occupy a central place in simulating real competition by comparing different companies and trying to bring the least performing to the level of the more efficient. In practice the implementation of quasi-competition faces numerous difficulties in the WSS sector. First, there is a big difference in the principle between dynamic competition in a real market and static comparisons between companies in a non-competitive market. Second, making meaningful comparisons between WSS providers is extremely difficult as there are many variables affecting the performance and the ambitions. Third, concentration on quasi-competition distracts attention from introducing real competition; for example the UK regulator forbids any more mergers and takeovers in the UK sector, as it wants to remain sufficient WSS providers to make a meaningful comparison.

5.6.6.1   The market to abstract raw water or discharge wastewater

WSS providers 'purchase' the right to abstract raw water from the natural environment and discharge wastewater to the natural environment. In return sometimes the WSS provider needs to pay a fee for abstraction or discharge to the 'owner' of the water resources. The owner of the water resources needs to accept that the service provider is implicating its natural water resources via these transactions. Water rights to abstract water and discharge wastewater are typically issued and controlled by the government.

Although opponents to the neo-liberal reform measures often confuse the debate by stating that the government has privatised the water resources itself, in reality such is mostly not applicable. The only arrangement in which such happens is through the creation of –so called- water markets (see Table 11). In this case privatisation is combined with liberalisation. Water markets are created when government have established a regime of legal tradable water rights.

Table 11 Neo-liberal manifestations for the market of the right to abstract or discharge water

Arrangements relevant for the market for the right to abstract or discharge water	Is it liberalization?	Is it PSI?	Is it privatisation?
Water markets	Definitely	No	Possibly

The existing water rights in the form of abstraction licenses and 'consents to discharge' could be subjected to some form of trading, time limitation and/or regulatory intervention. Introducing such type of competition in this market has proved to be very difficult. There are only a handful of examples in parts of California, Australia, Chile and Spain in which water rights are traded. The reasons are manifold but can be traced back to traditional reasons for market failure. Incumbent licence holders, especially the private service providers, have little incentive to trade the right to abstract water and discharge wastewater, since these are fundamental to their operations. Also informational bottlenecks proved problematic for organising such trades in a transparent way and setting acceptable maximum volumes for the trading system. The shifting of abstraction rights or discharge rights within a catchment area as a result of trades resulted in environmental externalities. In general, it could be argued that the risk of market failure is severe enough to restrict any wide scale adoption of liberalising the market for abstraction and discharge rights (Thobani, 1999).

5.6.6.2   The market for production inputs

Service providers rely on numerous suppliers of inputs in their execution of their tasks. They procure pipes, chemicals and many other types of operational materials. Energy needs to be purchased and different supporting services are procured to operate the infrastructure, such as leakage management, sludge disposal, and regular inspection of assets. Sometimes support is hired for activities related to customer liaison and account management, such as billing, meter reading, debt recovery, customer complaint and enquiry handling. More general supporting activities possibly procured are activities as regulatory affairs, human resource management, accounting and finance, research and development, information technology, planning, property management, public and employee relations, and vehicle management. Also supporting scientific activities such as sampling, measurement, monitoring, data analysis and presentation are sometimes procured. Also the employees themselves working in the service provider are 'purchased' from the labour market, which is typically a competitive market although the extent of competition is dependent on the scarcity of adequate employees, and the rigidity of the labour law. Furthermore, the service provider requires access to capital finance such as bonds, loans, grants, retained earnings and possibly share equity, to fund both infrastructure provision and to a lesser extent working capital. WSS service providers are continuously in need for financing of their investments. It is often argued that the discipline of the financial markets in the management and allocation of different risks is a vital ingredient in the potential success of private sector participation. Mainly banks are offering their services as suppliers of capital.

Each single input makes up in theory a separate (mostly competitive) sub-market. In these markets the regulatory interferences are trying to establish perfectly competitive markets (liberalisation) by establishing rules for tendering and regulating mergers between suppliers. In some cases the liberalisation goes hand-in-hand with types of privatisation and PSI (see Table 12).

Table 12 Neo-liberal manifestation for the market of inputs needed to produce

Arrangements relevant for the market for inputs (other than water) required to produce WSS services	Is it liberalization?	Is it PSI?	Is it privatisation?
Competition *in* the market	Definitely	No	No
Service contracts	Possibly	Definitely	No
BOT type of contracts[21]	Possibly	Definitely	Definitely

Competition *in* the market can be fostered as the markets to procure inputs are typically competitive, although the degree of competition varies between different product/service market segments. Regulation is in this respect aimed at trying to establish a perfectly competitive market by establishing rules for tendering and regulating mergers between suppliers.

---

[21] Although BOOT is included as a specific type arrangement, to emphasise the transfer of ownership to a private party, it is noted that often the temporarily ownership shift is merely a legal or financial construction.

PSI occurs in this market in case the public service provider establishes a partnership with a private party through types of service contract arrangements. The public service provider may decide to contract out some specific tasks to the private sector while keeping the responsibility for the overall management of the utility. These contracts are typically for short periods - from six months to two years - and take advantage of expertise of operators.

A specific type in which PSI is combined with privatisation is through so called BOT contracts (Build, Operate, and Transfer). In this case an arrangement is set up with the private supplier to construct part of it's' infrastructure, to operate it, and even to keep the temporary ownership during a specified (long) time.

5.6.6.3   The market of WSS services to end-users

In this market the service provider acts as 'supplier', while the 'end-users act as 'buyers'. WSS services can be further segmented into numerous sub-transactions. For instance, there is a sub-market for industrial water, for domestic water and for wastewater. Even there is a sub-market for making the water connections of customers. The relevant regulatory regime will be aimed at protecting customer interests, as the WSS service provider is (often) a monopolist. Consequently regulatory interferences in this market will focus on controlling tariff and service levels. The tariff can be controlled by: an organised competitive bid with formal contractual tariff resetting mechanisms (contract regulation), active regulatory control (rate of return or price regulation), passive regulatory control (investigation only if tariff increases proposed or complaints made by customers), or some form of self regulatory control (municipal rubber stamping of budgets or prices proposed by the service provider).

In practice, neo-liberal reform measures have little relevance for the market to provide WSS services to the end users. However, despite the dominant monopolistic nature there are some possibilities for end users to select among several suppliers. These possibilities mostly relate to competition *in* the market type of arrangements (see Table 13).

Table 13 Neo-liberal manifestations for the market of WSS services to end-users

Arrangements relevant for the market to fulfil the governmental mandate	Is it liberalization?	Is it PSI?	Is it privatisation?
Self supply	Definitely	No	Definitely
Competition *in* the market	Definitely	No	No

A first, mostly theoretical, possibility is to construct competing (or duplicate) networks. In view of the high costs associated with installing competitive networks, such option is regarded as not feasible. Another way to liberalise this market is by introducing so-called common carriage competition. In common carriage, the network owner allows access to a third party to supply some or all of its customers (Cowan, 1997). In order to allow several operators to use the same network a basic issue is the network access problem. The system is competitive only if new entrants have the possibility to use incumbent networks. The implementation of this form of competition entails a significant institutional framework to develop an efficient bulk supply and network access regime. Moreover, there are significant technical difficulties since mixing different waters can lead to water quality problems and potential health problems. Water is a perishable product, which might undergo undesirable changes during distribution and storage. The time taken for these changes to occur can be viewed as a form of shelf life. The mixing of water may influence this shelf life, and the additional residence time associated with the conveyance of waters over long distances cause the water to have exceeded its shelf life before delivery to the customer. Therefore, liability arrangements need to be carefully considered in advance (Aitman, 2001). England and Wales are the only regions that have made common carriage possible, since the Competition Act 1998. However, no such arrangements currently exist on a competitive basis.

A more feasible alternative might be to establish inset appointments. In this case, competition may occur in the form of competitive bidding to supply new groups of customers or for large customers. However, also in this case the incumbent provider has a large advantage in competing since it has established already its infrastructure and is tuned in to the local conditions. Moreover once these selected groups are connected, a monopolistic situation is established.

Finally, another type which might be mentioned, although under the header of 'privatisation' is self-supply. Self-supply occurs when one consumer (self-supply) or a group of consumers (co-operative supply) supply themselves rather than rely on an incumbent service provider. It can be identified as privatisation as a private party takes up all responsibilities and ownership of the service provision for a defined (small) area. In this case the local monopoly is broken by the presence of these (private) self-suppliers. A condition to make this option feasible is that water resources are accessible to customers.

5.6.6.4   Interrelations between the four markets

It is rather simplistic to make statements about liberalising or privatising the WSS sector since such measures only target a part (one market) of the entire sector. Whilst four main markets can be identified, it is important to understand the interrelations between the markets. Neither of the identified markets is operating in a vacuum. It might be that a neo-liberal reform measure is beneficial in terms of increased competition in one market, but has a negative net effect on the level of competition in the whole sector due to the impact it has on other markets. For example, if the municipal council decides to competitively tender the delegation of their service provision for an area (market 1), it will probably require greater competition in the market for abstraction and discharge (market 2) since such will increase the need to free up existing abstraction licenses and 'consents to discharge' through some form of trading, time limitation and/or regulatory intervention. Another example is when bidders are competing to get the mandate to serve temporarily a service region (market 1), the various input prices for operation, equipment supply, and cost of capital are effectively subsumed into a single all-in price (market 3). Hence, once the mandate is granted, especially the large multinational companies will be inclined to make use of their own supply chain companies to realise corporate profit margins, instead of organising competitive open tenders for these services. In this respect, the economic regulator OfWat (Office of Water Services) in England and Wales is closely monitoring the transactions between the private operators and associated suppliers of inputs. The linkage between the market 2 and 3 can currently be observed in Sweden (see Box 1).

The policy maker (and researcher) needs to be aware that the net effect of its reform measure on the whole of the sector can be entirely different compared to the impact they achieve in the target market. Each market has different regulatory demands and requires a tailor made set of regulatory tools to achieve varying regulatory objectives. Neo-liberal reform measures are manifesting themselves differently in the various markets. Each market has its own mixture of applicable reform measures, which are oftentimes combinations of privatisation, liberalisation and PSI. The fact that specific arrangements are in many cases combinations of liberalisation, privatisation and PSI is probably an important reason for the misconception that the liberalisation, privatisation and PSI are one and the same.

Box 2 Swedish illustration of market relationships

99% of the Swedish 289 municipalities are currently managing their WSS services provision either as a municipal administration or as a municipally owned limited company. Consequently, there is almost no competition related to fulfilling the governmental mandate to serve a dedicated service area. Only a handful of examples exist of municipalities organising competitive tendering for the right to serve a monopolised area (in 2002 there were six management contracts in Sweden). On the other hand the Swedish market to procure inputs required for production by the service providers is perceived as very competitive. As much as 70% of all goods and services needed to operate municipal water and wastewater works are brought on the market for suppliers in open competition. Most of these contracts are short-term contracts with renewal once a year or every second year. The argument made by pro-public representatives is that within the Swedish WSS sector the combined level of competition to transact the mandate to serve an area and the inputs to be procured by the service provider is higher compared to a country like France, in which the mandate to serve a region is to a high degree liberalised.

Source: Lannerstad (2003).

5.7   Synthesis of the Chapter

This Chapter completes the research design section. It adds to the previously defined research aim, the research methodological part. In this pursuit an analytical framework has been established in which several important choices were made. The analytical framework also guides the formulation of the main research question, accompanied by five sub questions for each level of analysis. Based on selecting strategy as the intermediate step between institutional changes and performance, the main research question for the thesis could be formulated. From the formulation of the research question, five relations could be identified which shape the analytical framework. Each of these five relations is to be researched, using different research techniques, e.g.:

1. The effect of neo-liberalism on the institutions in the WSS sector. This is studied through a case analysis of the Western European WSS sector (1).
2a. The effect of institutional changes on the strategy context of WSS providers. This is to be researched via a comparative case study in England & Wales and the Netherlands (2a).

2b. The effect of institutional changes on the strategy plans of WSS providers. This is to be researched via a single case study in the Netherlands Antilles (2b).

2c. The effect of institutional changes on the strategic actions of WSS providers, which is to be researched via a survey (2c), using the generic typologies of Miles & Snow.

3. The effect of strategic actions on performance of WSS providers (3). This last relation is to be researched by relating the outcome from relation (2c) to the performance of WSS providers derived from a country-by-country benchmark.

The choices made in the research methodology pertain to several aspects of the research. Each of these choices has a trade-off incorporated. On the one hand they provide the relevancy of the research and its added value to the existing body of literature; on the other hand they limit the research in its value and underline the caution that is to be taken in interpreting the research outcomes.

Limitations of the research relate to both methodological problems as conceptual problems. One methodological problem is the difficulty to select proper comparative sample groups, the statistical data analysis method selected, how to isolate the influence of non-institutional changes, and how to interpret the multi-dimensional construct of performance. A related problem is how to generalize the outcomes of the studies to the sector. The Chapter identifies within the complete WSS sector four markets, each again composed of numerous sub-markets. Purposely changing the institutions in one market by involving private parties or increasing competition may have effects in other markets, making the net effect for the sector very different. The identification of this problem is highly relevant for the research at hand as it implies that considerable modesty should be taken in formulating sector wide conclusions in case only one market is targeted by the research.

Other reasons are of a more conceptual nature. Especially, the fact that institutional changes do not have a direct effect on the performance of WSS providers, but they only influence performance through a change in conduct of WSS providers. Research trying to directly compare the performances of providers with different institutional characteristics ignores largely the intermediary variable of 'Conduct' in the terminology of the SCP paradigm. The identification of the disregard of the conduct of WSS providers when analysing neo-liberal effects is instrumental for the research at hand, as it opens a window for additional insight in how institutional changes may change the conduct of WSS providers, which then ultimately may change the performance.

In sum, this Chapter has set out the path to follow in the Analysis section, incorporating the lessons from the previous 'Introductory' section. It has identified the main research question, the analytical framework, its' main innovation and its limitations.

Section III: Analysis

Chapter 6: **Institutional Changes**

Chapter 7: **Institutions and the Strategic *Context* of Water Supply and Sanitation Providers**

Chapter 8: **Institutions and the Strategic *Plans* of Water Supply and Sanitation Providers**

Chapter 9: **Institutions and the Strategic *Actions* of Water Supply and Sanitation Providers**

Chapter 10: **Performance, Institutions and Strategic Actions**

Section III. Analysis

Chapter 6 Institutional Dynamics

This Chapter investigates the (neo-liberal) institutional dynamics in the WSS sector. A case study is conducted of the Western European WSS sector to identify the main elements. In this order the case study first assesses the driving forces towards adopting the neo-liberal agenda, followed by an identification of plausible futures of the Western European WSS sector.[22]

6.1   Introduction

The Introductory Section of the thesis already presented the adoption of the neo-liberal reform agenda as one of the major institutional changes in the WSS sector. This Chapter aims to provide further insight in the magnitude and rationale for neo-liberal institutional changes. In order to achieve this objective, a case study is conducted of the Western European WSS sector, in which first the drivers will be identified, second the institutional changes will be described, while thirdly plausible futures are presented.

6.2   Driving and resistance forces

The drivers and resistance forces with respect to adopting the neo-liberal agenda in Western Europe, are the first element of the case description (Van Dijk and Schouten, 2004b). The European WSS sectors are at a crossroad. Reports on the state of Europe's environment point at an alarming degradation of the Union's coastal, surface and ground waters. Moreover, new scientific evidence and public concerns have pressed for an amendment of public health standards for drinking and bathing waters (Kallis and Nijkamp, 1999). Public-private partnerships, models for subsidizing general interest obligations, the existence of publicly-owned systems and the role of regulatory bodies are some of the crucial issues of the debate. At the national and local level, pressures to liberalize the sector and to involve the private sector are not just due to the EU policy but to the overall restructuring of public systems: financial privatisation, the creation of mixed initiatives and partnerships are taking place in most countries, sometimes to improve efficiency and effectiveness, sometimes because of constraints on local public finance.

---

[22] Parts of this Chapter have been published in respectively:
- Schouten, M. and M.P. van Dijk, 2008. *Private Sector Involvement according to European water liberalisation scenarios.* In: International Journal of Water. Volume 4, No. 3/4, pp. 180-196. Inderscience Enterprises Ltd.
- Schouten, M. and M.P. van Dijk, 2007. Chapter *1. The European Water Supply and Sanitation Markets.* In: Water and Liberalisation: European Water Scenarios. Edited by M. Finger, J. Allouche and P. Luís-Manso. Published by IWA Publishing, London UK
- Schouten, M. and M.P. van Dijk, 2004a. *The European Waterscape of Management Structures and Liberalisation.* Water and Wastewater International Magazine, June 2004
- Schouten, M. and M.P. van Dijk, 2004b. *The dynamics of the European Water and Wastewater Sector.* Paper presentation on the International Specialty Conference on "Good Water Governance for People and Nature", September 2004, Dundee Scotland.

A myriad of drivers and constraints for adopting the neo-liberal agenda is identified across the Western European WSS service sector. It is the balance between these driving and resistance forces that will determine the propensity to change the existing institutions. Historically, this balancing act has resulted in predominantly direct public management of the European WSS sector. Currently, the majority of the Europeans (about 55 percent) receive WSS services from publicly owned operators, followed by privately owned operators (about 35 percent) and mixed owned operators (almost 10 percent). Pinsent Masons (2008) in its Water Yearbook forecasts for Europe that the amount of people served by private companies will substantially increase in the coming years (see Table below).

Table 14 Current and forecasted people served within Europe by private parties

	Population served by the private sector in 2008		Potential population served by the private sector by 2015		Potential population served by the private sector by 2025	
	Water	Sewerage	Water	Sewerage	Water	Sewerage
Austria	7%	0%	9%	14%	12%	17%
Belgium	3%	10%	3%	11%	3%	12%
Denmark	1%	0%	2%	2%	2%	2%
Finland	0%	1%	0%	2%	2%	2%
France	74%	55%	80%	71%	84%	76%
Germany	21%	18%	26%	29%	27%	31%
Greece	44%	37%	46%	45%	48%	48%
Ireland	1%	42%	21%	46%	19%	47%
Italy	41%	52%	47%	29%	52%	47%
Netherlands	0%	10%[23]	0%	11%	0%	11%
Norway	6%	0%	5%	10%	8%	12%
Portugal	25%	23%	56%	51%	61%	56%
Spain	43%	50%	63%	57%	64%	62%
Sweden	1%	1%	5%	5%	5%	5%
Switzerland	0%	0%	0%	0%	0%	0%
United Kingdom	88%	90%	94%	96%	94%	97%

Source: Pinsent Masons (2008).

Pinsent Masons made the above-presented projection based on historical growth rates of private sector involvement and an estimation of the influence of ongoing trends. However, in the Water Yearbook unfortunately the details of the calculation were not provided. In this respect the Euromarket project (Finger *et al.*, 2007) is complementary to the Yearbook, as this project has assessed which drivers and constraints can currently be identified within the European market. The nature of the identified drivers and constraints influences the extent and direction of change related to the adoption of the neo-liberal reform agenda. For example, the presence of powerful lobby groups for or against liberalisation may determine if the responsible government prefers to delegate the service provision to a public or a private entity.

The drivers and constraints can be categorized in three tiers (see Figure 11). The first tier of drivers and constraints is related to those parties most directly involved in the process of liberalisation: the incumbent service providers and the responsible entities for service provision. The second tier comprises the drivers and constraints related to those stakeholders that are more indirect involved with the service provision: the

---

[23] Explained largely by the contracting of the private party Delfluent by the Waterboard of Delfland for the Harnaschpolder project.

consumers, the press, the workers' unions and the (potentially interested) private parties. This group indirectly shapes the propensity and direction of liberalisation. The final tier of drivers and constraints relates not directly to a party, but relates to the context and the perception of liberalisation. The influence of the failure or success of flagship liberalisation projects in other sectors or countries cannot be underestimated: they shape the perceptions of stakeholders and provide the context for policy making.

Figure 11 Drivers and constraints to liberalising the European WSS sector

Source: Schouten and Van Dijk (2008).

6.2.1   The first tier of drivers and constraints

This tier of drivers and constraints relates to the stakeholders that are directly involved in the service provision. In this respect two main actors can be identified that shape the service provision: firstly the management entity that is executing the service provision, and secondly the responsible entity that bears the final responsibility for the service provision to the public.

6.2.1.1   The management entity

Since the management entity is the entity that is currently in charge of executing the service provision, its' influence on any changes to the service provision is quite significant. Several characteristics related to the incumbent service providers might trigger or discourage liberalisation. The most important drivers and constraints are depicted in Figure 13 and will be discussed. One of the important notions with respect to the interpretation of Figure 13 is that each of the characteristics can produce a different, or even an opposite effect, dependent on the local context. In the discussion of the drivers this situational dependence will be further clarified.

Figure 12 Liberalisation drivers and constraints related to the management entity

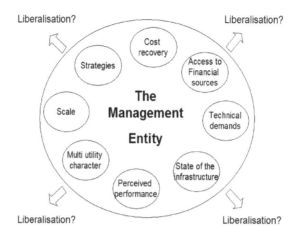

Source: Schouten and Van Dijk (2008).

- *The current scale of the incumbent management entity.* The strong urbanisation and the subsequent increasing proximity between municipalities have led a growing number of them to question the traditional mode of management in favour of larger management structures. This reasoning relies on the perceived financial benefits of economies of scale. One can now see in many European countries a strong growth of inter-municipal structures (like in Belgium, France, Germany and Italy) or of new regional structures (like in the Netherlands, Portugal, and England & Wales). For example in Sweden, Italy, Belgium, Switzerland and in the Netherlands it is believed that large incumbent service providers might be better able to maintain the required expertise themselves and avert interference from international operators. On the other hand, size also attracts private parties (Camdessus and Winpenny, 2003) since a minimum size is required to have a concession contract profitable in view of the transaction costs associated.

- *The current strategies of incumbent management entities towards market expansion.* For example in Greece the ambition of the incumbents to exploit emerging market opportunities in the WSS sector triggered partial privatisation of the two main WSS providers. An opposite effect can be witnessed for example in Scotland where worries about possible market entry by UK WSS providers searching for expansion, curtailed the liberalisation of the WSS market. Also this can be seen in Northern Ireland where employees of the incumbent management entities are lobbying to maintain the existing distribution of economic rents.

- *The current level of cost recovery of the management entity.* If there is a need to reach full cost recovery and make the required investment by attaining higher revenues this might trigger liberalisation as it did in parts of Switzerland. On the other hand, if there are limited opportunities for full cost recovery, the incumbent is a less attractive target in the eyes of private parties, as for example in Northern Ireland.

- *The multi-utility character of the management entity.* An existing integration between energy, gas and water in one management entity might blur the distinctions between the sector characteristics. As such, a multi-utility character makes the influence of the ongoing liberalisation in the other networking sectors stronger (especially the implementation of the EC Directive 96/92 on energy liberalisation) since the perception will be more dominant that what goes for one in-house sector should also go for the other. Such can be observed in Germany and Switzerland. Another element in this respect might be a push towards achieving economies of scope, as seen in the Netherlands where the Dutch water companies are eyeing towards the wastewater sector for integration, using the argument of water cycle management.

• *The present state of the infrastructure of the management entity*. The need to upgrade existing infrastructure, leads to foreseen increased financial pressure. Liberalisation might pave the way for enlarged access to funds. This is at hand in many countries such as Greece, Italy, the Republic of Ireland, Switzerland and the Dutch sanitation sector. If there is no real need to upgrade existing infrastructure, as in Scandinavia, there is also no need to search for other financial sources. Apart from the state of the infrastructure, the readiness of sound information available on the infrastructure can also be important. For example there is a lot of sound information on the assets of Scandinavian incumbent service providers, which is particularly interesting for private parties for making adequate value assessments.

• *The current (perception of) performance of the management entities*. If incumbent management entities are (believed to be) performing well as in Switzerland and the Netherlands, the urge to change might be lacking. However, if there is a strong desire to improve the operational efficiency of existing public management entities in combination with a perceived efficiency of the private sector, possibly due to high profile failures as in the drought in Greece, this might trigger liberalisation.

• *The present access to financial sources of the management entity*. If the public incumbents currently have the possibility to access cheap government loans, as in the Netherlands, this might hinder liberalisation since a private party would not have this access and so would incur increased financial costs.

• *Technical demands on the management entity*. If there is a situation of increasing complexity of WSS technology as for example in Greece, Italy, France and Germany due to environmental requirements on ground water abstraction and surface water discharge, there might be a need to involve private sector expertise and finance. Furthermore; the implementation of the European Directives for example in Belgium, Portugal and the Netherlands increased the required technical complexity and created a demand for private sector expertise and involvement.

6.2.1.2   The responsible entity

Characteristics related to the responsible entity that is currently in place are another major factor that needs be acknowledged. Several drivers and constraints (see Figure 13) are identified related to position, set up and abilities of the responsible entity.

Figure 13 Drivers and constraints related to the responsible entity

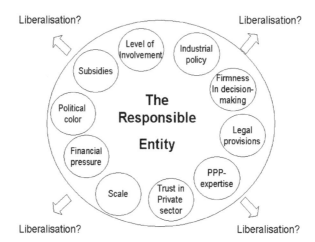

Source: Schouten and Van Dijk (2008).

- *The current level of involvement of the responsible entity on service provision.* If the incumbent management entity is highly dependent on government vagaries, this might seed the urge from the management entity to reduce the political vagaries in service provision by searching for alternative institutional arrangements. For example in the Scandinavian countries, the incumbent management entities are quite independent from government vagaries, and as such do not feel the urge to change. In other countries you see the opposite effect, such as in Northern Ireland and the Republic of Ireland, where the strong involvement of central government, either as a direct service provider or a major source of finance, prevents private sector involvement.

- *The current level of subsidies the responsible entity provides to the management entity.* If subsidies are in place, as for example in Austria, the government might be more inclined to liberalise the sector trying to relieve the financial burden of subsidizing.

- *The current political colour of the responsible entity.* In general one could say that a reigning liberal or right wing national/local political majority triggers liberalisation, while a reigning social democratic or more left wing national/local political majority is more resistant towards liberalising public services. However one could also say that political parties are acting within the national political dogmas. For example in Sweden and Belgium liberalisation is neither advocated by right wing nor left wing politicians, while in Greece both political sides choose liberalisation.

- *The current financial pressure on the responsible entity.* For example, conforming and/or joining the Economic and Monetary Union and the Stability Pact criteria pressured the Greek, Spanish and Belgium government to reduce public debt and triggered the partial privatisation of WSS providers to generate additional financial resources. Other examples are the Republic of Ireland and Portugal, in which a decline in EC Regional Funds led to a search for alternative sources of finance. Also in the Swedish, Swiss and German municipal governments, the desire to reduce the financial burden on the public budget triggered the involvement of private parties.

- *The current scale of the responsible entity.* A large number of small municipalities acting as responsible entities may encourage associations, which may then be attractive for private sector involvement. For example in France, Belgium, Portugal and the Republic of Ireland, PSI is used to create multi-municipal structures and hence gain economies of scale.

- *The current level of trust the responsible entity has in the private sector.* For example, the long history of interaction and hence trust between private and public sectors in France provides a good basis for partnerships, whilst in the Republic of Ireland there is little confidence and trust in the private sector possibly due to a lack of historical precedence. Historically, the service providers, as part of a municipal administration, had low degrees of autonomy being subject to municipal political decision makers. Currently there is a clear trend of governments to grant higher degrees of autonomy to the service providers. Such can be seen in cases when governments are setting up separate public companies for service providers that operate at arms' length from the municipal or regional public administration. This trend towards more autonomy can be due to the increasing need for professionalization of these sectors and the application of New Public Management concepts (Schwartz, 2006).

- *The current level of PSI-expertise of the responsible entity.* If regulatory expertise or general expertise about PSI is lacking, the hesitance of introducing private sector involvement increases. An example of a country in which the lack of PSI expertise may constrain further liberalisation is Sweden. However in case the expertise is available or if it can be readily obtained from neighbours, as in Scotland, this characteristic may be a driver for future liberalisation.

- *The firmness of the responsible entity in decision-making.* If politicians are unable or unwilling to take the blame for an increase in tariffs they might be inclined to use the private sector as an instrument to undertake a politically difficult task. Conversely if politicians remain firm in their decision-making: as in Scandinavia, Denmark and Luxembourg, and they are willing and able to introduce tariff increases and to undertake considerable efforts to comply with the Water Framework Directive this may not be the case.

- *The current legal provisions in place that encourage PSI.* For example in the Republic of Ireland, Spain, Switzerland and some German Länder; arrangements for private sector involvement in the WSS sector are under current consideration, yet in other countries, such as the Netherlands, Italy and Sweden, there are still major legal constraints to PSI. On the other hand the current "pipeline developments" at EU level indicate that the European Commission is looking for ways to heighten the exposure of the European WSS sector to competitive forces. If these developments materialise in legislation it would definitely trigger liberalisation processes.

- *The industrial policy of the responsible entity.* For example in Belgium, the industrial policy of the government triggered an up-scaling of regional companies to allow them to develop activities abroad. Politicians hope that the current investment in the wastewater sector will benefit regional construction companies and will contribute to capacity building of the construction companies. Also the German Bundestag initiated a so-called sustainable water management policy that targets reaching optimal efficiency gains through modernisation of the present system. The constituent policy elements will affect the organisation of the sector and may lead to shifts in the present structure as well as to increased competition and private involvement. In contrast, Spanish politicians use the argument of public responsibility for sustainable aquatic ecosystems to enhance the need for public management responsibilities in the WSS sector.

6.2.2   The second tier of drivers and constraints

The second tier of drivers and constraints with respect to liberalisation of the WSS sector, are those related to the stakeholders and pressure groups. The following groups can be identified:

• *Press*. If the press are extensively involved, as in Sweden, this obstructs the liberalisation process since some politicians might be inclined to play the public sentiment on this controversial topic. If the media does not feature the liberalisation process, such as in Greece, implementing private sector involvement might turn out easier.
• *The private sector*. Lobbying by private companies that want to enter the WSS sector might be a driver for liberalising the sector, as is the case in France, Spain and partly in Switzerland.
• *Workers' unions*. The fear of job losses due to private sector involvement in the WSS sector might trigger them to lobby against liberalising the sector, such as in Northern Ireland.
• *Consumers*. If the consumers pay a relatively low price for WSS services, as in Switzerland, or if they feel they might loose from any proposed restructuring, as in the Netherlands, Spain, England and Wales, the consumers might be inclined to block changes through public consultations. A specific group of consumers to exercise considerable influence are the large industrial users. They are lobbying for their interests if they feel they will gain from restructuring and increase in competition. These organisations lobby both at the European level (like the International Federation of Industrial Energy Consumers) and at the national level (like the Dutch VEMW Association for Energy, Environment and Water and the German VIK Verband der Industriellen Energie- und Kraftwirtschaft). Corruption scandals in some countries have led consumers to be more suspicious about private sector involvement. By increasing the transparency of the sector, stakeholders will be better able to understand the way the sector is managed and to have a better control over price mechanisms. Governments are also seeking to increase the transparency as to create possibilities for better control, regulation and possibly involvement of private parties. Also the service providers themselves are calling for more transparency as to anticipate regulatory impositions and as part of a public relation strategy.

6.2.3   The third tier of drivers and constraints

Aside from the direct and indirect incumbent parties within the WSS sector, an important factor that influences possible liberalisation is the experience of liberalisation elsewhere, either in other parts of the country, other countries or other sectors. Even before liberalisation is on the agenda, one of the first things to look for is an assessment of the successes and failures apparent elsewhere, and to use these learning experiences for their own situation. Regional evidence that PSI in neighbouring utilities in the WSS sector is working well might convince others to also enter a PSI process, as in Sweden or the pioneering Dutch DBFO contract in Delfland. Furthermore the experiences abroad of successful PSI can be influential; for example the efficiency of English WSS providers is shown in Scotland as a success and something to aim for. In contrast, the same case of England and Wales is reported in the Swedish media as a poor example of liberalising WSS services. If international benchmarking surfaces a poor performance, as with the Greek public sector, this might trigger private sector involvement. Also liberalisation experiences in other (networking) sectors might provide either a driving force or a resistance force towards liberalising the WSS sector. Local evidence that PSI works well in others sectors, as in Scotland, definitely triggers PSI in the WSS sector. The institutional changes in the WSS sector received an important push from technological change in the telecommunications industry. Government gradually dismantled state monopolies in telecommunications and opened the door to private providers. This demonstrated that privatisation and competition could deliver improved infrastructure services (Braadbaart, 2005). On the other hand when liberalisation in other sectors turns out negatively, as the unbundling of the electricity sector in Northern Ireland or the railways in the Netherlands, opponents use it as an argument against liberalisation of the WSS sector. Furthermore possible policy development with regard to liberalisation may be highly influenced by the perceptions and experiences relating to the corruption of the private sector, like for example the PSI contracts in Grenoble and Milan (Hall, 2007).

6.3   The current European WSS sector

The second element of the case description concerns the institutional changes in the Western European WSS sector (Schouten and Van Dijk, 2008). In this respect an overview is provided of how the sector is organised and the direction to which the investigated driving and resistance forces are leading.

The simplicity and clarity of the water delivery process seems to be in stark contrast with the complexity and diversity in which the sector is organized. The WSS sector is characterized by very different types of providers, ranging from small non-autonomous municipal services, whose main goal is the provision of a public service in a geographical limited area, to private trans-national corporations (TNCs) pertaining to profit maximisation and to the expansion of operations worldwide. From a stable and merely municipal service, the sector is turned into a more dynamic and heterogeneous sector incorporating a multiplicity of institutional arrangements. The private sector is playing a more dominant role and the service providers are becoming more autonomous, bigger and subject to competition and calls for transparency. Policy makers at all levels, from the European Commission to the local municipalities are realizing this shift by implementing numerous sector reforms measures. Governments are modifying legislation, setting up independent regulatory bodies, organising tendering procedures for delegation contracts, encouraging mergers of WSS providers and are exploring new ways of managing and controlling the provision of WSS services. Each individual country finds itself currently in a different institutional context, and it would be valuable to investigate such in relation to the previously identified driving and resistance forces to adopt the neo-liberal reform agenda.

The European WSS sector is known for its' diversity and complexity. In the case study, analysis of the institutional context of the WSS sector is conducted at pan-European and EU Member State specific level. The objective of the analysis is not to assess in depth the market structure characteristics but to be able to provide a broad overview of the European WSS providers.

To provide an assessment of the most dominant institutional arrangement currently present in each Western European country and the dynamics based on indicative trends in the country is made use of a matrix constructed by two axes, identifying (i) direct or delegated management, and (ii) private or public management. These two elements were previously discussed in Chapter 2. The matrix indicates four types of dominant market structures on the axes of direct or delegated management and public or private management, with the arrow indicating the increased degree of separation between the Principal and the Agent. The common factors that differentiate the institutions are similar to the ones used in the Eureau studies, being: (1) The degree of organisational autonomy of the Agent; (2) The degree of autonomy of the Agent in setting of the tariffs for the end-user; (3) The degree of autonomy of the Agent to access funds from other sources than the Principal; (4) The degree of ownership of the assets for producing and distributing water and wastewater by the Agent, and (5) The degree of regulation and control enforced by the Principal upon the Agent.

The dominant institutional context in each Member State will be assessed and positioned within the matrix indicating its relative position at pan-European level. Based on this positioning Member States are clustered that more or less find themselves in the same situation, e.g. (1) the direct public management cluster, (2) the cluster in change towards delegated public management, (3) the cluster in change towards delegated private management, (4) the direct private management cluster. Investigating the institutional arrangement in the various Western European countries, the follow overview matrix can be presented:

Figure 14 Overview matrix management structures European WSS sector

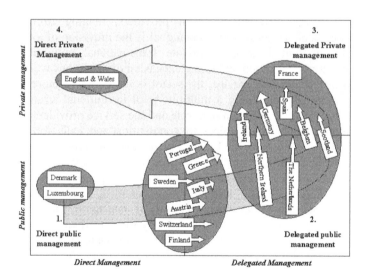

It needs to be noted that within each of those groups large differences can appear. Of course, any methodology employed involves a certain measure of subjectivity in the choice criteria and the overview really represents a broad-brush approach to the subject. Especially in view of the great diversity of market structures, even within the same country, the positioning of a country within this matrix is a major simplification of a much more complex situation in reality.

6.3.1   Stable direct public management cluster

This is a cluster of two countries that dominantly organised its' WSS provision in the form of direct public management, e.g. Denmark and Luxembourg. Although there are some minor indications of trends towards private sector involvement and greater autonomy of the Agents, still the constraints towards change are large.

Direct public management was historically the most adopted market structure that European Principals chose to organise the management of WSS services in. In many countries this model has undergone severe changes, but for a select cluster of countries (Luxembourg and Denmark) the constraints for change were too strong to increase the degree of separation between the Principal and the Agent. The main constraints, which are more or less recognisable in both countries, are the strong record in providing good service levels of the incumbents. Therefore the urge to change is absent. If change is planned, opposition mobilises in the form of labour unions, incumbent service providers and customer organisations. Moreover, the financial pressure upon the municipalities is not very strong. The municipalities have strong revenue raising powers and the current infrastructure is performing well and for the near future not up for renewal.

Denmark[24]. For Denmark the WSS sector is strongly in the hands of the public sector, except from the private co-operatives that own and manage their own private wells of which there are about 9,000 serving one third of the population mainly in the rural areas. Municipalities assume the roles of Principal and Agent (in the form of a municipal department) in the provision of WSS services. As such the degree of separation between the Principal and the Agent is very limited. The market structure is strongly direct public management and the major trend identified is the increase in the scale of operations. Over the past 22 years, the number of common utilities bas been reduced by 55 utilities per year on a straight-line basis.

Luxembourg[25]. Under the law of 1843, WSS provision is the responsibility of the 118 municipalities (District Councils). The service is not delegated although municipalities are allowed to form inter-municipal associations (Syndicats). Currently there are seven Syndicats for water and eleven for sewerage. These Syndicats limit themselves to producing and feeding drinking water to the municipal tanks, while water distribution remains the responsibility of the municipalities themselves.

6.3.2   Cluster in change to delegated public management

Within this cluster of seven countries, Agents are getting more and more autonomy. The cluster is marked by a change effort from direct public management towards delegated public management, sometimes even with some private sector involvement in the form of minority shareholding. Within this cluster, the Principal wants to extend the degree of separation with the Agent, but still holds the intention to keep it in public hands. In some countries experiments with delegated private management are undertaken but they are still on a small scale and very much in its infancy. Variations within this cluster are large. For example in Portugal and Switzerland a small number of concession contracts with private parties are signed. Also in some countries minority shareholding of private parties is in place. For example in Greece the two

---

[24] The data on Denmark is based on: Eureau (1997); OECD (1998) Stockholm Environmental Institute (2003); European Commission (1997); DWWA (2001); http://www.danva.dk/sw114.asp

[25] The data on Luxembourg is based on: Aluseau and PriceWaterhouseCoopers (2003); Eureau (1993, 1997).

major WSS providers of Athens and Thessaloniki established private minority shareholding and in Sweden one municipality even sold all of its' shares to the private sector.

The main drivers for change towards delegated public management are the increased financial pressure due to investments needed for rehabilitation and upgrading. Especially for instance in Greece the increasing focus on environmental problems of water resource infrastructure development and the need to improve the operating efficient of existing public sector bodies, resulted in decreasing the link between the Principal and the Agent. In Sweden, the smaller rural municipalities are rapidly merging their water and wastewater activities to attain efficiency improvements and find a solution for the lack of investments. Another driver for change is the failure in some instances of direct public management, for example in Greece when they coped with the drought between 1989 and 1993. In some countries another factor is the desire to create large players that can participate in the emerging WSS sector in Europe, especially South East Europe.

Austria[26]. The Austrian 1.900 municipalities engage themselves in 99% of the cases both as Principal and as Agent in providing WSS services. In Austrian rural areas sometimes associations of consumers take upon themselves this role. At the moment in Austria, there are 4.000 water systems, but only 200 exceed the 500 inhabitants supplied. These 4.000 water systems are managed by the municipalities and are supplying 87% of the population (the remainder relies on self-supply). In Austria there is a trend noticeable that long distance water and wastewater distribution is aggregated between local suppliers. At the moment, there are 150 local authority associations in Austria, created to provide regional WSS services. The involvement of the private sector is currently limited. In Austria private parties are allowed to hold a minority share in the WSS companies. It is estimated that private parties manage only 5% of the WSS services. Also in sanitation only 1% of the population is served by pilot projects involving private sector involvement. The debate on private sector involvement in Austria is much more targeted at the sanitation sector, mostly for ethic concerns. Main driver for the debate on private sector involvement is the increased financial pressure upon the sector. Between 1993 and 2001 investments in WSS infrastructure amounted to about 105 billion EURO of which almost 37 billion EURO was financed through public fund transfer. In recent years public funding has been decreased to 0.2 billion EURO. This decrease of public financing, together with an estimated investment between 2000 and 2012 of 12.7 billion EURO, pushed for private sector involvement. In fact, a recent document by PriceWaterhouseCoopers for the Minister of Environment recommends to introduce some form of concession contracts in the Austrian WSS sector.

Switzerland. The role of the Principal in Switzerland is assumed by the 26 cantons and is normally delegated to the municipalities (around 3,000 in 1997) although the extent of delegation will vary from Canton to Canton. Municipalities usually manage these services directly. A major trend noticeable in Switzerland is the move from direct municipal provision to the establishment of autonomous public entities operating under public law. Participation of the private sector remains limited. No liberalisation is evident and not envisaged in the near future. In smaller municipalities network maintenance tends to be contracted to the private sector. There appears to be a

---

[26] The data on Austria is based on: Rossmann (2001); Hansen *et al.* (2003)

distinct difference between WSS sectors when the question of delegation arises. For water, municipalities may choose to provide the service directly or to delegate it to a third party. For sanitation, the municipality must manage the service either directly or through a syndicate of municipalities.

Greece[27]. The Greek municipalities are acting as Principals. Outside of the two major cities, Athens and Thessaloniki, they are also assuming the role of Agent. Mostly the municipal Agents organise themselves either as municipal owned public utility corporations (for 30% of the population) or as municipal departments (covering 20% of the populations). For the two major cities, Athens and Thessaloniki, two state owned companies have recently been restructured and partly privatised. The state floated 25.5% of the Thessaloniki Company and 28% of the Athens Company. These two Companies manage about half of the Greek population. The minority participation of the private sector as shareholders is part of government policy to reduce state involvement. Further privatisation may follow in view of the desired reduction of public deficit.

Italy[28]. The Italian municipalities assume both the role of Principal and Agent in 82% of the cases. Recently there has been a development ongoing towards the establishment of public companies operating at arms' length from the municipality. The establishment of the so-called OMA's[29] is indicative in this respect, as a single operator manages the OMA. OMA's are originated from the decision of municipal companies to join together and sell only a minority of shares. In Italy no legal obligation exists for delegation also to private sector parties. The Galli law allows specifically management alternatives as formal privatisation, legal privatisation and delegation. But only in two cases there is full privatisation, covering 8% of the population in 1996, but this Figure is increasing.

Portugal[30]. The 275 Portuguese municipalities assume the role of Principal. Agents are predominantly public, i.e. direct public management in 224 municipalities, municipalized services in 46 municipalities, and 3 public companies. Only six municipalities decided to delegate this task to private companies under concession contracts. Despite the privatisation decree of 1993 only a few concessions have been signed since then. The Portuguese WSS market remains relatively closed to private participation (only 10% of the population is served by operators with private sector participation). The PEASAAR[31] indicates strategic options to solve problems of Portuguese water systems. These are the adoption of integrated solutions on a territorial basis and an operational basis, and the transformation of systems in corporate structure. The document considers supra-municipal systems as the best way to attain efficiency in service provision and fulfil the lack of investment needed. The Water National Plan from 2001 confirms this solution as a way to release the State and take advantage of investment capacity as well as the know-how of the private sector.

---

[27] The data on Greece is based on Tsagarakis *et al.* (2001).

[28] The data on Italy is based on: Comitato di Vigilanza sull'oso delle risorse idriche (2003); Banca Intesa (2003); Holzwarth and Kraemer (2000); Massarutto (2000).

[29] Ambito territoriale ottimale, i.e. Optimal Management Area is a syndicate of municipalities.

[30] The data on Portugal is based on: Baptitsta *et al.* (2003); INAG (2002); MAOT (2000); Martins and Fortunato (2002); DREE – Mission Economique de l'Ambassade de France (2003).

[31] PEASAAR is the strategic program for water industry in Portugal, presented in 2000 by the Ministry of Environment.

Sweden[32]. Although a new trend in Sweden is to establish Public Limited Companies, still the local community administrates the majority of the WSS provision. In 1995 there were 17 Public Limited Companies and this amount has already doubled five years later in 2002. The majority are multi-utility companies with activities in fields as electricity, waste, roadwork and district heating. Only a minor number are strict water companies (5) as Stockholm Vatten AB. Only 7 communities provide the services through management contracts with private parties. Private Sector involvement in the WSS sector is a new phenomenon and it started in the smaller municipalities. There is currently one fully privatised municipality and seven management contracts with private parties, of which one is concluded with a foreign party (being Veolia).

Finland[33]. In Finland 446 municipalities assume the role of Principal for providing WSS services. The municipal council determines the operating area of public water works and sewage works and then takes care of services to that area. Over 90% of total water supplied is via municipal water works. The remainder is supplied via "private" associations, as cooperatives, partnerships and stock companies (owned by the users). Usually WSS provision is part of the municipality's technical organisation but there is a trend of establishing delegated public management structures going hand in hand with the trend towards the merging of municipal WSS providers. In 1988 there were 487 municipal water works, in the year 2001 there were 452, while currently there are 442 municipal companies, and 20 mergers are currently being discussed. In Finland several forms of delegated public management exist: there are five joint municipal authorities that have some autonomy, there are thirteen inter-municipal wholesale water supply companies, and there is one joint stock company for water and six joint stock companies for sewerage. More joint stock companies are planned: six for water and eight for water and sewage (mainly in the biggest cities), but progress has been slow. According to the Water Service Act from 2001 the municipal water works have to keep accounts separate from the main municipal accounts. Most public water works operate on a commercial though non-profit making basis. Typically in the larger municipalities the Agent operates as a utility (e.g. Tampere, Turku) or as a limited liability water companies (e.g. Helsinki Water, Oulu). Also the Water Services Act describes that all contracts and charges relating to water supply will be governed by private law. There is no apparent move in Finland to involving the private sector, since there is in Finland no obvious financial pressure due to the fact that the investment is largely made already in the 1970s and the 1980s. The track record of the municipalities is strong and consumers have a good opinion about the service levels.

6.3.3   Cluster in change to delegated private management

Delegated private management is becoming increasingly important in Europe. Especially the so-called French model attracts a lot of attention. Quite a large group of European countries is in change towards delegated private management, although the group is quite diverse. Some of the countries within this cluster, such as France, have a long tradition of delegation to private parties, for others, such as the Netherlands, this process is still in its infancy. The Netherlands were historically set up as direct public management, they transformed in the last decades of the previous century in delegated public management and now initiatives mainly in the wastewater sector

---

[32] The data on Sweden is based on Lannerstad (2003).
[33] The data on Finland is based on Katko (2000).

seem to indicate that delegated private management is considered. For example, in Northern Ireland there is definitely a push towards a larger involvement of the private sector but progress in this respect is slow.

Major drivers found in these countries pushing towards delegated private management are that involving the private sector in procuring infrastructure projects is increasingly favoured. Especially DBO (Design Build Operate) contracts are frequently used, as for example in Ireland. The private sector involvement softens the financial pressure the public sector is coping with and the belief is that it enhances efficiency improvements. Also successful local and international PSI experiences trigger others to initiate PSI arrangements.

France[34]. The 36,000 French municipalities are acting as Principal for WSS provision. The municipalities are free to choose between several modalities to manage WSS provision. There are two alternatives implemented:
1. Direct public management, where the municipality takes the management of services in its own hands.
2. Delegated private management (75% of the drinking water services and 35% of the sewerage services), where municipalities sub contract their duties to private companies. Two contracts are currently in use: affermage and concession. Private parties are strongly involved. Currently there are three main private actors active in the WSS sector, being: Generale des Eaux, Lyonnaise des Eaux and SAUR, which have respectively 8.000, 5.000 and 7.000 municipal contracts, serving 26, 14 and 10 million people in France. About 50 smaller private companies operate at local and regional level.

Ireland[35]. In Ireland, the local authorities act as Principals. The management structure is based on original 26 counties of Ireland. Acting as Agent within each county there is one (or more) Sanitary Authority that is charged with the provision for WSS services. A limited number of authorities agreed to manage an inter-municipal service. Remarkable for Ireland is the abolition of domestic water tariffs, as conceded in 1996. Non-domestic customers are charged but not for the full costs incurred. The Minister for Environment stated in the National Development Plan for 2000 to 2006, that he expects PSI to increase. Especially DBO (Design Build Operate) contracts are common in Ireland. At the end of 2002 over 70 projects approved to advance as DBO schemes and a further 60 were being examined. The majority (i.e. over 75%) of projects in procurement relate to sewage management.

Northern Ireland. The Secretary of State for Northern Ireland assumes the role of Principal for the WSS sector at central government level since 1973. The Water Service, an executive agency established in 1996, acts as Agent, managing WSS services for 99% of the population for water and 83% of the population for sewerage. The Water Service was established by the conservative government to act as an interim step to possible privatisation and act as vanguard for Private Finance Initiatives. Following devolution of certain powers to the Northern Ireland Assembly in 1999 statutory responsibility has resided with the Department for Regional Development. With devolution on hold this has returned to central government. PSI is mentioned as a possibility for the future in the Strategic Investment Programme

---

[34] The data on France is based on: Barraque *et al.* (2000); Délégation a l'aménagement et au développement durable du territoire (2003); Haut conseil du secteur public (1999); Holzwarth and Kraemer (2000); IFEN (2003); Boyer and Garcia (2002).
[35] The data on Ireland is based on WS Atkins (2000).

although currently there is only one DBFO in place. Just as in Ireland there are no domestic tariffs in Northern Ireland. Revenues are covered by the public expenditure system, although the Central government commented that water charging would be put in place by 2006. Still the Central government feels it is committed to reform the WSS provision in Northern Ireland. A major consultation on water reform was undertaken in the summer of 2003.

Germany[36]. Acting as Principal in the German WSS sector are the municipalities. Historically the German municipality is playing a major role in the WSS sector but with the increase of technology and the rising costs numerous other institutional settings have been developed. The WSS sector scaled up because of the expensive water treatment, which needed a high level of operational competence and often the creation of larger units. The optimum sizes of these units were often beyond the range suitable for individual enterprises. Besides scaling up, also a trend towards horizontal integration is noticeable in the establishment of integrated municipal companies, providing electricity, gas, water, etcetera. Besides municipal companies, also medium sized enterprises play an important role nowadays. In Germany, municipalities cannot delegate the responsibility for the sewerage while they can for drinking water supply. In Germany the debate on private sector involvement progressed, mainly in the drinking water sector. There are some noticeable examples of this, being the concession contract for Berlin and the involvement in some Eastern Germany cities as Rostock.

Belgium[37]. Acting as Principals in the Belgium WSS sector are mainly the three regional administrations (Flanders, Wallonia, Brussels). Next also municipalities assume sometimes the role of Principal. Most dominantly in Belgium, Agents are organised as Regional companies (2) or inter-municipal associations (26), of which some are mixed companies. The two regional companies (VMW and SWDE) serve about half of the population. Next there are some Municipal companies (17) and some concessions (5) granted to private parties. Municipal companies are managing the sewage collection in Belgium. Two of these municipal companies are involving the private sector in the form of a cross-border lease contract. Sewerage treatment is in Wallonia organised through inter municipal associations. Only in Flanders and Brussels another form is dominant for sewerage treatment. Flanders and Brussels established two regional companies for wastewater treatment, in the form of mixed companies, serving all of Flanders and Brussels. In all of Belgium a strong trend towards increased private sector involvement is noticeable. The involvement of the private party Aquafin NV as the Regional Company for the wastewater treatment services in Flanders, the several inter-municipal associations that attract private parties as minority shareholders, the recent BOOT contract between the Region of Brussels and private companies and the five concluded concession contracts for the cities of Gent, Oostende, Vevriers, Tournai and Malmedy are worth mentioning in this respect.

Scotland[38]. Scottish ministers are acting as the Principal for WSS services. Scottish Water assumes the role of Agent since it was established in 2002 by the merger of

---

[36] The data on Germany is based on: Francisco (1998); Rudolph (2001); OECD (1998); Umwelt Bundes Amt for Humanity and Environment (2002); http://www.destatis.de/basis/e/umw/umwtab3.htm.
[37] The data on Belgium is based on: Alaerts (1995); Eureau, (1997); Euromarket (2005); Varone and Aubin (2002).
[38] The data on Scotland is based on: Water Industry Commissioner for Scotland (2001); Scottish Executive Environment Group (2003); Scottish Water (2003).

three regional public WSS providers. Scottish Water manages the WSS services to 98.5% of the population for drinking water and 92.5% of the population for sanitation. In sanitation Scottish Water relies heavily on Public Private Partnership Financing. There are nine separate wastewater PSI projects. They currently process over 80% of the total wastewater of Scotland and account for virtually all of the wastewater treatment in non-rural areas. No more large-scale PFI projects are currently envisaged. Focus is now on network maintenance and renewal. Scottish Water recently announced the launch of a new group Scottish Water Solutions to deliver 70% of its £ 1.8 billion investment programme by 2006. Scottish Water will own 51% of the new group. The joint venture company will be responsible for all remaining asset work prescribed, capital works will no longer be competitively tendered.

Spain[39]. In Spain the 8.000 municipal authorities are acting as Principals for WSS provision. Cities larger than 20,000 inhabitants often delegate the services to private companies. Private participation now reaches around 50% of the supplied population. The main group is Aguas de Barcelona that supplies around 25% of the population (836 municipalities), and then comes FCC (with Veolia participation) serving 18% of the population, and Bouygues Saur serving 7% of the population. In Spain the traditional model of private sector involvement is the concession one. However the shared public-private company model has also developed, now representing 15% of delegated private management. Private companies are minority shareholders in this model. Municipal administrations are managing the WSS provision in smaller towns, which is relatively more easily to manage; representing 32% of the population. A third arrangement in Spain is that public commonwealths or regionally organized municipalities manage the drinking water provision for larger areas (13.5%). Sewerage networks and treatment are mostly managed directly by municipalities, but also here the major cities delegate the sewerage service provision to private companies through concession contracts. Aguas de Barcelona serves 396 municipalities (or 15 million people), FCC serves almost 10 million people and Bouygues Saur serves about 750,000 people. Municipalities or local companies manage the rest of the sewerage services. One can observe a clear trend toward more private sector involvement in the field of water supply management due to the fact that municipalities are confronted with high investments to conform to EU directives and are faced with exhausted public funds, and growing technical complexity of infrastructures.

The Netherlands[40]. The Dutch market structure is fragmented. Three separate markets exist each composed of a different structure; (1) the drinking water market, (2) the sewerage collection market, and (3) the sewerage treatment market. The drinking water market is dominated by Public Limited Companies acting as Agents on behalf of municipalities. Still two direct public management companies exist but all other companies have merged into 14 limited liability companies. It is foreseen that also these 14 companies will be merging and even a smaller number of companies will remain. For the Dutch drinking water sector there seems to be no indication of a move towards delegated private management. The second market is the market for sewerage collection. Here municipalities assume the role of Principal and Agent, although many

[39] The data on Spain is based on: Asociación Española de Agua y Saneamiento (2000); Asociación Española de Abastecimientos de Agua y Saneamiento (2003); http://www.aeas.es.

[40] The data on the Netherlands is based on: Dalhuisen (2003); Rioned (2002); Unie van Waterschappen (1999); Van den Berg and Van de Reyt (1996); Vewin (2001).

types of delegation towards public as well as private parties are common. The third separate market is the market for wastewater treatment that is dominated by the water boards. Especially in this third market an interesting development is noticed with the first DBFO contract for the biggest wastewater treatment plant in Europe (Delfland). The Dutch sanitation sector is closely monitoring this case and if successful it is likely others will follow.

6.3.4   *Stable direct private management cluster*

This cluster is a group of one. Direct private management is still uniquely applied in England and Wales, where the private party has acquired the complete ownership and management of the WSS provision. This is the most extreme form of separation between the Principal and the Agent. It needs to be noted that an increased degree of separation does not imply there is also more competition in this cluster. The English WSS providers were given without competition in 1989 25-year concessions. The WSS providers enjoy the right to 25 years' notice before their concession can be submitted to competition (Hall, 2003). The central government, as the Principal, is involved for regulatory issues. Elsewhere in Europe there seem to be few drivers and many constraints to also consider entering such a management structure. Most of the constraints deal with the resistance to place goods of 'public nature' such as WSS services completely in the hands of the private sector.

England and Wales. Acting as Agents in England and Wales are ten large water and sewerage companies (WASCs) and 13 small companies that only supply water (WoCs). Next there is 1 very small water supply company. The majority of companies are owned by holding companies that are quoted on European stock markets. The National government, as Principal, institutionalised several regulators (such as OfWat, the Drinking Water Inspectorate and the Environmental Agency) to control the management and operations of the Agents. The trend in which England and Wales are moving is to enhance the competitive elements in the WSS sector, for example by introducing common carriage.

6.4   The future Western European WSS sector

The third element, this Chapter aims to highlight with respect to the policy background, is the future development of the European WSS sector. In this respect plausible futures of the European WSS sector are explored by identifying a set of liberalisation scenarios for the period to 2020. The objective is not to predict or forecast the future, but to create a window on which current actions and developments can be reflected. The scenarios presented were developed in the 3-year EU financed Euromarket project.

A scenario can help to see what the future will be like and (via the storyline) how/why these futures occur. Scenarios represent what is plausible, not necessarily what is either desirable (i.e. normative) or probable (trend based). Scenarios are purposely challenging, being designed to help us confront the assumptions we are making about the present and future. Scenarios are valuable because they stimulate questions rather than because they provide answers. Effectively the scenarios describe different paths (via a consistent set of events, trends and actor strategies) that lead to alternative futures. A scenario is therefore composed of two separate elements: the End State, which describes the situation at a particular future point in time, and the Storyline, which connects the current state to the end state in a logical manner. Hence a scenario is both a description of the future and how we get to that future.

The art of scenario development is to reduce a large range of possibilities to a handful of plausible directions that together contain the most relevant uncertainty dimensions. The Euromarket project[41] used competition processes as the basis for the development of the scenarios. The EU scenarios are not solely based on the competition modalities. Management modalities such as ownership structures (public or private), and/or the degree of separation between the responsible and management entity (direct or delegated) are also relevant. Based on the competition modalities and the management modalities five plausible scenarios were defined each emphasizing a different market and competition process within the framework. The scenarios prepared for the future development of the European WSS market range from a more explicit role for the government as provider or regulator to a situation with more private sector involvement. It will not be useful to impose one model for all countries. Rather the different models may converge towards a model with more private sector involvement and closer regulation. Each of the scenarios has strong and weak points and all in all, the results depend on the institutional and organisational responsibilities defined in terms of regulatory choices, risk sharing, and level of co-operation and understanding between the different entities involved in the sector. Therefore, there is no point in recommending a particular scenario but rather to outline critical governance issues posed by each of them and to address the various solutions for these issues.

Figure 10 was used to construct the scenarios. Recall from this Figure that essentially the WSS sector may be subdivided in 4 markets, in which each market can again be further subdivided. In each of these markets different types of competition can be identified. To construct the scenarios this framework is leading. Each scenario emphasises one particular type of competition in one of the markets becoming dominant in the WSS sector. The dominant competition process is used to determine the fundamental nature of the European liberalisation scenarios. The scenarios are graphically represented in Figure 15 below:

---

[41] For more information on the Euromarket project including all the freely downloadable reports, you are advised to visit the website: http://www.epfl.ch/mir/euromarket

Figure 15 European scenarios

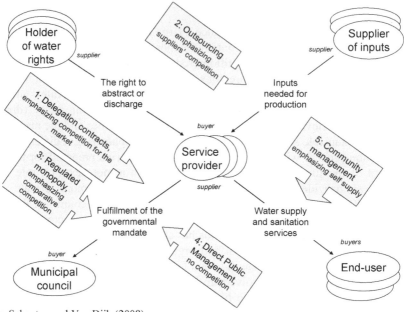

Source: Schouten and Van Dijk (2008).

6.4.1   Scenario 1: delegation contracts

The EU wide scenario is based on the assumption that from 2005 to 2009 transnational companies are the major actors influencing the EU policy towards more delegation contracts. Several driving forces underlie this scenario. First, European transnational companies are retrenching back to Europe because of the heavy losses incurred in low and middle-income countries. They lobby actively for an EU directive on competition for the market. Second, new EU entrants and some existing Member States face heavy investment outlays in order to comply with EU quality standards on WSS service provision. Combined with recent GATS negotiations, an EU directive proposal is launched in 2008 that imposes an obligation on local/regional authorities responsible for WSS management to tender, in order to promote competition for the market. This directive will be agreed in 2009 (before the change in European Commission and Parliament) and must be applied by 2012 by all Member States.

The scenario is divided into two variants, depending mainly on the length of the tendering period obliged by the EC directive.

Under the first variant, from 2010 onwards, NGOs, left wing parties and local/regional authorities gradually challenge this influence and lobby for tighter regulation of operators. Parallel to this movement, the European Commission is also very interested in developing evaluation of performance for Services of General Economic Interest (SGEI) and plans to launch an "EU evaluation of performance of SGEI". It defends the establishment of independent regulators in the WSS sector at national level that control and diffuse performance indicators of different operators. According to the story line, this growing social demand for more ex post regulation of operators managing delegated contracts will lead to a new EU directive in 2020. This directive concerns the establishment of independent regulatory authorities to ensure that operators respect public service obligations (availability, continuity, affordability). The end state of this first variant is one in which the EU WSS sector is dominated by delegated contracts combined with strong regulation. There is competition *for* the market every 10 to 15 years. Ex ante regulation (in order to choose the more efficient operator) is established by competitive bidding. Concession or lease (affermage) contracts are the most common contractual form, awarded mainly to transnational companies. Markets are not unbundled, with private companies, both transnational companies and smaller national companies, managing a vertically integrated service. Tariffs are agreed for long periods of time according to rules for adjustment based on indexation formulas. The operator is responsible for tariff collection and bears the risk of non-payment. The responsible entity in the public sector remains the legal owner of the original assets, with the responsibility for investment and maintenance depending on the nature of the contractual arrangement. The scale of the responsible entity is not necessarily synonymous with the scale of service and may be responsible for several services on different networks. Although monopoly power remains there will be ex post regulation by independent regulatory authorities, which use performance indicators to control the water price and quality of service and to ensure respect for public service obligations.

Under the second variant, transnational companies remain the major actors influencing EU policy towards more delegation contracts throughout the period to 2020, with a much more reduced role for political parties, NGOs, and local/regional authorities in comparison with the first variant. There is some delay in introducing legislation for five-year contracting in Member States where public management or total privatisation is predominant. Nonetheless, the new directive is finally applied. The end state of this second variant is characterized by delegated contracts, combined with extreme competition resulting. The high level of competition is due to an EU directive requiring responsible authorities to introduce open tenders every five years, with contracts awarded solely on the basis of least cost. This leads to market domination by an oligopoly of the largest private European WSS providers with the gradual disappearance of direct public sector management. Stringent environmental standards and policies must be applied in all countries, and be integrated in the terms of the contract and invitations to tender. The regular five-year tendering procedure is considered to be sufficient to promote competition. Therefore, there is no need for specific regulation other than national and European competition authorities.

6.4.2   Scenario 2: outsourcing

This scenario is based on the simple question, "What happens if no dramatic or critical events take place, overall water use remains more or less stable, and the current trend toward greater efficiency continues". This 'quiet' scenario is compatible with the wide variety of different institutional arrangements that are already found in EU Member States because all of them include 'outsourcing' to a greater or lesser extent.

The main economic drivers for this scenario are the long term underlying trends that are already present in the first decade of the 21st century. Foremost among these is a strong drive towards greater efficiency in service delivery. More investments are required because of the modernization of existing WSS systems, compliance with EU directives, the shift in supply from ground water to high-cost surface water, increase in scale that requires investment in physical infrastructure, investment in sanitation and the need to make infrastructure less vulnerable to terrorist attacks. In response to this demand for extra funding for investment, there is a continuous political pressure to enhance the efficiency of the sector and to offer public services at a lower, cost-related tariff. This pressure for greater efficiency encourages European operators to increasingly outsource some of their tasks to sub-contractors. These tasks range from short-term (1-5 years) well-specified contracts for activities such as network repairing, billing, management, as well as longer-term contracts such as DBFO, BOTs, and partnering for infrastructure specific maintenance/construction. Unlike under the delegation contracts that feature in Scenario 1, the revenue risk here is not transferred to the winning bidder for these outsourcing contracts. There are four driving forces behind the expansion of outsourcing: benchmarking, the early successes of member states that have adopted extensive forms of outsourcing, new EU legislation on outsourcing that lowers the threshold values beyond which contracts have to be awarded through public tendering (this extends outsourcing from the area of customer services to that of management), and finally the increase in demand for specialized expertise. The expansion of outsourcing, together with accession of the new member states to the EU, leads to internationalisation of the sector. Low cost and highly competent companies from Central and Eastern Europe move rapidly into the water outsourcing market. The dynamics of outsourcing begins to shape the sector. Subcontractors develop services for sectors other than the WSS sector, so becoming multi-utility subcontractors. According to this scenario, from 2015-2020 citizen trust in regulatory bodies in a number of member states is put to the test as a result of disappointing results in terms of efficiency and effectiveness of the regulatory framework. There is a growing demand for specialized knowledge in order to overcome the information asymmetry between the tendering operators and the specialized sub-contractors and consultancies. A European support body for public procurement procedures is established in order to compare performance among sub-contractors. This body develops a standardized 'model' contract for outsourcing that gains worldwide attention after widespread coverage in the international media. This legal product is exported and heavily propagated by the World Bank in developing countries.

6.4.3   Scenario 3: regulated monopoly

This scenario is characterized by benchmarking as the key competition process in the monopoly markets. This takes two forms. High-powered benchmarking with centralized regulation takes place where there are private monopolies subject to a strong external and independent regulating authority at the central level. The regulatory authorities determine the tariffs, budgets, prices and investments that companies may charge or carry out. Medium-powered benchmarking with decentralized regulation takes place in those countries, where the organizational structure of the sector is characterized by strong municipal influence. The publication of benchmarking information exerts public pressure on companies.

Under this scenario, in those Member States where private monopolies have been installed they are viewed as the only possible remedy to the problems that were inherent in the preceding arrangements. In those Member States where the municipalities' influence prevails, liberalization is not viewed as desirable for the WSS sector and there is a general antipathy toward full private sector ownership, as long as a good service is provided at reasonable prices. Assets are owned by the operators, who range from private companies to highly autonomous municipal undertakings that work at supra-municipal level. They tend to provide all WSS services and manage all assets. There have been no EU liberalization directives and

EU activity has been restricted to the promotion of benchmarking initiatives. There is full cost recovery, including environmental and resource costs in some countries. There are no direct subsidies, but indirect subsidies and regional cross-subsidies remain.

The EU storyline for this scenario is composed of a series of driving forces that influence the overall path followed to the final End State. The most important of these are: public attitudes against liberalization; extreme pressure on municipal financial and managerial) resources leading to the need for divestiture of WSS services; higher tariffs due to environmental and health concerns as well as moves toward full cost recovery; EU policy promoting the need for modernization and compulsory benchmarking. From 2010 onwards, four complementary driving forces continue to drive the process of institutional reform in the European WSS sector. These are: EU reviews of cost recovery and drinking water quality that promote the need for further transparency and that encourage further benchmarking activity; implementation of the Water Framework Directive (WFD) underpinning major structural changes in scale and integration of WSS operations; a rapid growth in consumer power stimulated by abuses of monopoly positions with demands for greater economic regulation. From 2015 positive experiences of countries that have adopted independent benchmarking/regulation together with changes in the global macro-environment reinforce the pace of institutional reform. Positive international experiences of benchmarking/regulation become well known across the EU. Climate Change impacts both directly and indirectly on the WSS sector. Severe droughts and floods appear to overwhelm the system in a number of countries that had not yet implemented major water reforms. Along with the WFD (see above), climate change encouraged the formation of integrated WSS operators based on water basins. This reorganization enables greater economies of scale and scope.

6.4.4   Scenario 4: direct public management

This scenario is characterized by the absence of competition in/for the customer market or for various service inputs. The operator is typically a non-autonomous local public WSS services body under the direct control of the municipality, which is the sole provider of integrated WSS services to the community. Although some operations are outsourced, contracting out is restricted to large turnkey infrastructure provision and to the high technology domain. There is no independent regulatory authority. Instead each operator acts as a regulator in its region.

The storyline is illustrated by two pathways that converge towards the Direct Public Management end state: one of status quo and the other of convergence after an external shock. Under the first pathway, good quality and affordable WSS services leads to a gradual evolution towards an improved and strengthened direct public management by a series of adaptive innovations through the introduction of New Public Management tools (NPM) and the implementation of the Water Framework Directive (WFD). Under the second pathway, external events (e.g. contamination accident, natural disaster) or voluntary actions of the various actors (e.g. corruption scandal) lead to a strong reaction from civil society, successfully pressing public authorities to turn to direct public management.

6.4.5   Scenario 5: community management

This scenario is characterized by the participation of the community in the provision of WSS sector in the following ways: the sector is organized in voluntary organizations (i.e. user co-operatives); customers own water assets or can contribute to WSS sector management through representation in water company boards; the WSS sector is a responsibility of water management associations formed by landowners, private enterprise or public corporations. Decisions concerning WSS services management are decentralized by transferring responsibilities to communities. There is no competition in/for the customer market or for various service inputs and contracting out is generally restricted to infrastructure provision or for technological expertise demanding tasks. The community retains the ownership of the infrastructure and takes strategic decisions concerning the level of service and financing. The community may be involved in the day-to-day operation and maintenance or it can delegate this task or some other aspects to a professional body.

The storyline towards the End State begins with technological innovation (e.g. reverse osmosis, decentralized sanitation systems) and company self-supply from wells, spreading the Community Management model initially in rural and dispersed populations, and to new tourism areas where the extension of good WSS services presents high integration costs for public and private operators alike. Elsewhere, limited financial sources boosted the search for cost effective solutions, both in public or private management systems. From 2010, some private and public owned companies decide to transfer ownership of WSS services to citizens. Encouraged by a growing mistrust of both public authorities and private companies, some local communities begin to express a strong preference for the development of decentralized systems in order to avoid the construction of new systems (thus reducing investments in WSS infrastructure). Strategic decision-making in the WSS sector is strongly influenced by a wider societal crisis of individualism, under which people feel growing satisfaction in public activities rather than only in private consumption. By 2020 the Community Management model ceases to be a residual form of the WSS sector management in areas not served by centralized WSS services and is generalized, with widespread involvement of users at local level, through ownership or participation in decision making. The implementation of the Water Framework Directive and of the Aarhus Convention requires increased involvement of citizens in environmental decision-making and facilitates their active participation. The spread of the Community Management model is spurred by growing diseconomies of scale in large and growing urban areas that make decentralized systems with new technologies both cheaper and more efficient.

6.5   Synthesis of the Chapter

The Chapter aimed to assess the impact the adoption of the neo-liberal agenda had, and will have on the WSS sector. In this regard, a case study description of the Western European WSS sector was conducted. The case study description was composed of three elements: the drivers for change, the institutional changes in the WSS sector, and the plausible futures.

The case study shows that the European WSS sector is at cross roads. A myriad of drivers and constraints is identified for adopting the neo-liberal agenda. By making a distinction in three tiers of driving forces the case study aims to bring further understanding that it is the balance between these driving forces that will determine the propensity of change. The forces presented in the first tier of driving forces exercise the most direct influence. This first tier makes up characteristics related to both the management and the responsibility entity. The second tier composes the indirect stakeholders, like the press, the consumers, the private sector, and the unions, which may also influence the direction of change with respect to liberalisation. The last tier of forces is not directly related to any of the stakeholders in the WSS sector, but refers to the perception of liberalizing other sectors. Many examples provide the decisive influence of such perceptions on the viability to liberalise also the WSS sector. However, obviously the drivers identified may be country specific. The analysis of the institutional changes sheds more light on the context specificity of the identified drivers and constraints.

The identified drivers and constraints have historically created a shattered landscape of institutions all over Western Europe. Four clusters of countries are identified based on the direction they are heading with respect to possible liberalisation. The first cluster is labelled as the stable direct public management cluster. This is only a small group of countries (e.g. Denmark and Luxembourg) in which currently the service provision is directly provided by public entities and in which drivers for change are largely absent. The second group comprises seven countries. These are the countries in which also the sector is dominantly characterized by direct public provision, but drivers for change towards a more delegated public model can be observed. The third group is the largest group. These are the countries that are currently dominantly organised as the delegated public management model, but in which tendencies can be observed to delegate responsibilities to private parties. The final cluster is composed of one country only. This is the unique case of England & Wales, which is labelled as the stable direct private management cluster. Hence, from the clustering of countries it can be concluded that the vast majority of countries finds itself in a transition towards more delegation either to autonomous public parties, or to private parties.

The final element of the case study attempts to construct plausible futures for the European WSS sector. Based on the identified drivers and constraints 5 story lines and end states in the year 2020 are envisaged. What becomes apparent from the identified scenarios is that in all cases elements from the neo-liberal agenda become increasingly important. Four out of the five scenarios envisage higher levels of private sector involvement in some way or another. One scenario is quite different from the others (e.g. the Community Management scenario). However, even in this scenario one could interpret the spirit of neo-liberalism in which competing private parties (communities) take on the ownership and operations of localised WSS systems.

In sum, the case study shows the relevancy of neo-liberal institutional changes now and in the future. The institutions in the WSS sector have changed, and will continue to change. Policy makers are, and are expected to keep on, weighing the implications of adopting a neo-liberal agenda in the WSS sector. In this effort they can use all the help from the scientific community on the possible outcomes.

Section III. Analysis

Chapter 7 Institutions and the strategic context of WSS providers

This Chapter aims to identify to which extent different institutions restrict WSS providers in shaping strategies of their choice. Hence, a comparison is made in a case study of the managerial discretion allowed in a neo-liberal institutional context and a public institutional context[42].

7.1   Introduction

This Chapter addresses the first level of analysis of the implications of neo-liberal reform measures on the WSS sector, e.g. the implications on the strategic context. The central research question this Chapters 7 is concerned with is: to what extent can differences in institutions explain differences in the strategic context of WSS providers. An analysis of the regulatory institutions is an essential condition to study strategies of WSS providers. Researchers have noted that different regulation may be conducive to certain strategic actions (Desarbo *et al.*, 2005). Strategic context is interpreted as the managerial discretion of WSS providers to pursue strategies of their choice. It is hypothesised that institutions make a difference for the strategic context of WSS providers.

The Chapter is structured in the following manner. A case description is provided from the UK and the Netherlands. The case study is divided in two separated but interrelated parts. The first part describes the current administrative regulatory framework that governs the WSS sector. In this respect the perspective from the regulator is central. In the second part of the case description the perspective shifts towards the regulated. This part tries to identify how the managerial discretion compares of WSS providers subject to the regulatory regimes. The comparison uses the 5 strategic components of Boyne and Walker (2004) to structure the analysis. For every one of the strategic components an analysis is made to what extent the regulatory regimes provide opportunities or set constraints. A case study comparing the impositions and opportunities provided by the regulatory regimes in the UK and the Netherlands is instrumental in that respect. England & Wales was selected as the case to represent the private companies. In England & Wales currently 23 private companies are responsible for WSS service provision, of which ten companies

---

[42] Parts from this Chapter have been published as:
- Schouten, M. and M.P. van Dijk, 2005. *Regulatory impositions posed upon strategic actions of publicly and privately owned water companies, in respectively the Netherlands and England and Wales.* Paper presentation on the International Symposium on "Competition and stakes in the regulation of services of general interest. Feedback of the last twenty years", September 2005, Paris, France
- Schouten, M., Van Dijk, M.P., Swami, K. and M. Kooij, 2003. Chapter 4: Country Report The Netherlands. In Deliverable 4: Analysis of the Legislation and Emerging Regulation at the EU Country Level. Work package 4 (phase 2) Euromarket. December 2003.
- Schouten, M. and M.P. van Dijk, forthcoming (accepted May 2008). *Regulation and comparative discretion of publicly and privately owned water companies in the Netherlands, England and Wales.* Water Policy. IWA Publishing

provide WSS services, while thirteen are Water only Companies (WoCs)[43]. The comparative case from the Netherlands features as an example of a traditional publicly managed model. In the Netherlands 14 Public Limited Companies assume the responsibility for supplying drinking water services; municipalities handle the sewerage collection services, while the water boards are responsible for the treatment of wastewater. Both countries are relatively homogenous and have comparable economic, social and cultural conditions. The Chapter finally makes some concluding remarks.

7.2   The regulatory institutions along the water chain components

In the description of the regulatory institutions, the process of WSS provision is divided into three: resource access, water production & distribution, and sewerage collection & treatment. In each sub-process, legislation in force, objectives, instruments, target groups and actors of implementation are briefly introduced. Furthermore, the policy rationale and effects (outputs and outcomes) of the different parts are identified. Annex 5 provides a complete overview of both regulatory regimes along this sub division.

7.2.1   Accessing the resource
According to the UK Water Resources Act (1991) it is an offence to abstract water or discharge polluting material, solid waste, effluents or any matter likely to hinder the water's flow, unless permitted by the Environment Agency. The UK waters are seen as a "*shared resource*", as the UK government is managing rather than owning UK waters (Zabel *et al.*, 1998). The UK regulatory regime is aimed to prevent over-abstraction and to improve control over the environmental effects of water abstractions; to ensure a fair and efficient allocation of water between competing local demands; and to contribute to maintaining and enhancing the quality of water dependent environments.

Similarly to the UK situation, water resources in the Netherlands are considered as a "res nullius" (no property). Water resources cannot be owned, not even by the state, because it is considered to be public (Kuks, 2003). The Dutch government manages and controls the surface as well as the available groundwater through the Water Management Act and the Ground Water Act.

A number of instruments have been used in both countries to achieve these objectives and will now be discussed.

Abstraction licences are generally granted in the UK based on the demonstrated right of access of the applicant for at least one year to the land where the abstraction will take place. Licence applications then need to be brought to the attention of those likely to be affected by them. In case a licensed abstraction causes damage or loss to anyone, the person has the right to seek financial compensation from the abstractor. The Environment Agency charges abstraction licence holders in order to fund costs incurred in ensuring that water resources are managed effectively. Charges consist of an application fee as well as an annual fee that is based on volume abstracted and is linked to local water resources. With a view to preventing or controlling the entry of

---

[43] Water only Companies supply drinking water to certain areas within the regions covered by the major Water and Sewerage Companies but are not involved in wastewater collection, treatment and disposal. As far as drinking water supply is concerned, the duties of water only and water and sewerage companies are the same.

any polluting matter into controlled waters, areas can be designated as water protection zones, with activities in such zones being subject to specific restrictions (Water Resources Act 1991). Through the Environmentally Sensitive Areas Scheme, programmes are operated that aim to encourage good farming practices within these areas. Seventy-seven rivers in England and Wales are classified as Sites of Special Scientific Interest, which represents the UK's top conservation status. Over abstraction of water represents serious environmental problems for sensitive rivers in England and Wales. The designation of protection zones has increased considerably in recent years in an attempt to better preserve vulnerable areas and to comply with EU regulations. Despite this trend, only an estimated 80% of the length of England and Wales' 77 "Sights of Specific Scientific Interest" (SSSI) rivers are considered to be in a 'good' or 'very good' chemical condition. Many of the rivers classified as SSSI have been affected by diffuse pollution. High phosphate levels were found in 54% of the rivers in England and Wales, and excessive nitrate concentrations were found in 29% (Defra, 2003). A number of planning instruments (e.g. Regional Development Plans) have been made operational at different levels (e.g. Environment Agency and its regional offices). Specific regulations apply to manure spreading fertiliser use and pesticides. For pesticides, a strict approval process and Codes of Practices on their use have to be followed. A voluntary package of measures to reduce the environmental damage caused by pesticide use was implemented by the industry and other stakeholders (as of April 2001).

The Groundwater Act of the Netherlands prescribes that Provinces are qualified to define regulation and to select the different extractors. Large-scale extractors are obligated to request a permit while small extractors (less than 10 cubic meters per hour) only have to be registered. The Water Boards coordinate the surface water permissions if the source does not belong to the "State waters". In the case of the rivers the Rhine and the Meuse, which are considered to be state waters, Rijkswaterstaat is responsible. Water abstraction charges apply only to groundwater extraction in the Netherlands. Two taxes are being charged since 1995; the first one is being charged by the provinces and varies between € 0.01 and € 0.08 per cubic meter. The purpose of this tax is to contribute towards the cost of research for developing groundwater policy plans. The second tax is charged by the Ministry of Finance and is not hypothecated to the environmental purposes. Drinking water providers pay about € 0.34 per cubic meter under this second tax. There is a rebate of € 0.28 per cubic meter in case surface water is injected into the groundwater prior to extraction. The implementation of groundwater charges has had little effect on the drinking water providers to use surface water instead of ground water. The amount of groundwater extracted as a part of all water extracted for drinking water production has seen a minor drop and is about 62% (Versteegh and Biesebeek, 2004). The objective, a 10% reduction of the depleted areas in the year 2000 has not been reached. According to the Third Environmental Policy Plan a 40% reduction of the depleted area in 2010 compared to 1985 can still be achieved but requires an intensive effort from both Provinces and water boards. Because currently 10% of the total area of the country is affected by a permanent lowering of groundwater tables the first outcomes can be considered weak (Kuks, 2003).

Comparing the Dutch and the UK regulatory regimes, they are not very different. In both cases the WSS providers do not own the water resources and are dependent on temporary permissions to abstract and discharge. Also in both cases the WSS providers need to pay a volumetric abstraction charge to the government. The main difference between the two cases is that UK WSS providers communicate with one state agency (the Environment Agency), while Dutch WSS providers need to interrelate with several institutions (provinces, Ministry of Finance and water boards) to arrange the right to access water resources. Table 15 depicts the legislation in force in the Netherlands and England & Wales.

Table 15 Legislation in force on resource access

The Netherlands	England & Wales
Ground Water Act (1981)	The Water Act (2003)
Soil Protection Act (1986)	The Water Resources Act (1963/1991)
Water Management Act (1989)	Natural Heritage (Scotland) Act (1991)

Source: Euromarket (2005).

7.2.2   Producing and distributing drinking water

The regulatory regime in both countries is aiming to ensure the efficient provision of good quality drinking water and compliance with national and EU regulations at affordable prices for consumers through quality standards and controls, economic regulation (charges), monitoring and transparent public information policies.

A number of instruments have been put in place in order to achieve these objectives in the area of water production and distribution, which will be further addressed in the following sections.

In England and Wales, only one in five households has a metered water supply that allows for bills to reflect the amount of water used[44]. As charges based on actual use are not feasible without metering, most drinking water in England and Wales remains charged as a rate on an old property tax base. As of 1 April 2000, all household customers are entitled to have a meter installed free of charge and the number of households with a meter has increased since then. Drinking water providers have a duty under the Water Industry Act 1991 to supply water that is wholesome at the time of supply, i.e. when water passes from the water company's pipe into the consumer's pipe. "Wholesomeness" is defined by reference to standards and other requirements set out in the Water Supply Regulations 1989 and 2000. Drinking water providers are required to submit to the Authorities programmes of work designed to secure compliance with the new and the revised standards. Drinking water providers are themselves responsible for ensuring the quality of their supplies through adequate monitoring. This 'self-monitoring' role is, however, subject to supervision by local authorities and the Drinking Water Inspectorate, which conducts continuous technical audits to ensure that drinking water providers are meeting all their regulatory obligations[45]. Drinking water providers are obliged to publish the results of their monitoring activities and make all results of regulatory sampling available to the general public via their public record. Information must be provided to the interested public free of charge. Furthermore, drinking water providers have to produce an annual report on drinking water quality for the local authorities in their supply area. Section 70 of the Water Industry Act 1991 makes it a criminal offence for a drinking water provider to supply water that is unsuited for human consumption. Section 18 of the Act requires the Authorities to take enforcement action for any breach of

---

[44] Per company, the proportion of domestic customers metered varies considerably and ranges from 3.3% (Portsmouth Water) to 53.7% (Tendring Hundred Water) (Ofwat, 2002b).
[45] Reports of these audits, where relevant, are available on the web site at: www.dwi.gov.uk.

wholesomeness standards, monitoring and treatment, and/or records and information requirements of the Regulations. However, enforcement action is not taken in certain circumstances[46]. Around 99% of the population of the England and Wales is currently connected to the supply network (Eurostat, 2007). As OfWat reports in its tariff and charges report 2002, the charges set by the water companies for the provision of their services broadly correspond to the costs of providing these services, for metered and non-metered customers alike (OfWat, 2002b). With regard to drinking water standards, 99.8% of sample tests in England and Wales in 2002 complied with the relevant standards[47]. It is worth noticing that the distribution system in England and Wales is characterised by a high leakage rate (with an average rate of 22% in 2002). In order to foster reductions in leakage rates, OfWat set the medium term objective in 1997 of achieving a so-called Economic Level of Leakage (ELL) "which is the point at which the cost of reducing leakage is the same as the value of water saved" (Defra, 2003).

The Dutch policy objectives for the drinking water distribution system are to provide sustainable drinking water distribution services to everyone and as a result improve the public health conditions. Other objectives are to operate the service on a cost-recovery basis and to stimulate efficient water use. Two types of prescriptive legislation can be distinguished in the Dutch case. The first is related to the quality of the service provided and the other to the organisation of the drinking water supply service. The required quality and pressure are described in the Water Supply Act and updated in the Drinking Water Decision from 2001. According to the Water supply Act, drinking water providers are responsible to provide drinking water of the required quality. The act directs the drinking water providers to supply wholesome drinking water to the users in quantities and pressures required to protect public health. If there are problems in the distribution that might influence the quality of the drinking water delivered to the customers, drinking water companies are obligated to inform the users. The Water Supply Act stipulates that plans made out by the provincial councils may require the water companies to supply drinking water in bulk to one or more water companies at prices that cover all the costs. Even when supplying water to its customers the water companies are expected to charge tariffs at cost recovery levels. The government used to supervise the compliance to these requirements by way of annual inspections, but these have now been discontinued and the supervision is now based on trust complemented by periodic inspection. The drinking water companies are obliged to collect and hand over the information asked for by the inspectors from the Ministry of Environment. Self-regulation started in the year of 1989 with annual performance reports to improve efficiency. In 1997 the VEWIN started a Benchmark study, which has been executed three times by now. Different indicators related to water quality, services, environment and finance are collected and compared for more then 85% of the Dutch water companies. The benchmark study is used to increase the transparency of the performance of the companies and to provide an instrument, which can be used to improve the company's processes[48]. The drinking water service in the Netherlands is of a very good quality

---

[46] Enforcement action is not taken if the breach is: (i) deemed to be trivial; (ii) unlikely to recur; (iii) the water company has taken immediate remedial action to prevent a recurrence; (iv) or the water company has submitted a legally-binding programme of work to achieve compliance within an acceptable time scale.
[47] For more detailed results, see DWI at http://www.dwi.gov.uk/pubs/annrep02/mainindex.htm

[48] Benchmark studies 'Water in Zicht' by VEWIN (2004 and 2007).

and almost never fails. The consumption per person is quite low compared to other Western European countries[49].

Comparing the Dutch and the UK cases, large similarities can be noted in the regulatory regimes. In both cases the drinking water providers are required to provide wholesome water, and monitoring is done by themselves. One of the striking observations is that in the UK both the percentage of connections without a meter and the percentage of Unaccounted for Water are much higher than in the Netherlands. Below an overview is provided of the legislation in force related to producing and distributing water.

Table 16 Legislation in force on water production and distribution

The Netherlands	England & Wales
Water Supply Act (1957) and revisions (2000) and (2002) Drinking water decision (2001, quality norms) Policy Plan (for 30 years (VROM) and for 10 years (VEWIN)	Water Industry Act (1991) Water Industry Act (1999) Water Supply (Water Quality) Regulations (1989) Water Supply (Water Quality) Regulations (2000) Private Water Supplies Regulations (1991)

Source: Euromarket (2005).

7.2.3   Collecting and treating wastewater

About 94% of the UK population is connected to sewers leading to sewage treatment plants, of which 11% are of preliminary treatment; 13% are with mechanical treatment, i.e. primary level; 55% have biological treatment, i.e. secondary level; and 21% are of advanced treatment, i.e. tertiary level (Defra, 2002). Compared to other European countries, this rate of tertiary treatment is exceptionally low (Eurostat, 2007). Most of the remaining population is served by small private treatment works, cesspits or septic tanks (Defra, 2002). The regulatory regime is aimed to secure that the wastewater collection and treatment is adequately controlled and cost recovery is established.

Almost all of the Dutch population (98%) is connected to the sewerage system. Problems occur because of the condition and the existing capacity of the sewerage system. Because of leakages the wastewater collected by the sewerage system infiltrates into the groundwater. In many municipalities there is insufficient capacity to transport all the wastewater. Problems seem to grow because of the increase of inhabited surface area and the heavy rainfalls. Full cost recovery as described in the EU's Water Framework Directive of 2000 has not been achieved in the case of sewerage. Sixty percent of the wastewater treatment is not capable of removing nitrate according the European Nitrate Directive.

The main instruments regulating the collection and treatment of wastewater are sewerage charges, pollution levies and discharge licenses. These will now be discussed in the following sections.

Sewerage charges are mostly derived on the basis of standardised property values used for taxation purposes. The remaining population pays volume-based sewerage charges. Overall, OfWat demands that charges should broadly relate to the costs of

---

[49] OECD (1999) reports consumption levels per capita in the Netherlands of 130 lpcd, in Germany of 129 lpcd, and in England of 141 lpcd.

providing the service, for metered and non-metered customers alike. The individual WSS providers are, however, relatively free in setting tariffs and can charge different regions or classes of customers differently as long as they do not discriminate. All discharges to controlled waters require the granting of a discharge consent. There are two main types of discharge consents, namely numeric and descriptive consents. Discharges, which have the greatest potential to affect the quality of the receiving water, have numeric concentration limits attached to their consents. These limits may apply to an individual substance or groups of substances. Through the Environment Act 1995, it is allowed to transfer a discharge consent to another person who proposes to carry on the discharge in place of the existing holder. If a discharge consent has been granted for discharges into controlled waters, a discharge charge is levied from the Environment Agency. This charge is meant to fully recover the costs encountered by the Environment Agency in fulfilling its pollution control function. The scheme includes both an application charge and an annual charge that is based on three factors, namely volume, content and receiving waters.

In the Netherlands, pollution levies are considered an effective incentive to reduce wastewater discharge (Bressers *et al.*, 1993). During the period 1980-1991 the industrial and communal organic discharges to surface waters dropped by 51% (Buckland and Zabel, 1998). The point sources of pollution are well taken care of in the Netherlands; the diffused sources of pollution are a major problem. The Dutch regulatory regime aims to protect the environment through operating wastewater efficiently and taking care of reducing negative impacts to the environment. In the Netherlands all environmental legislation is incorporated in two acts: the Environment Management Act and the Pollution of Surface Waters Act. The objective of these acts is to promote the purification and prevent the degradation of the surface water. For sewerage these objectives translate into avoiding the dilution of wastewater in the sewerage system by minimising infiltration, prevent groundwater pollution by leakage of wastewater from sewers and also to connect every establishment and other wastewater producers to a treatment unit wherever possible in effect trying to minimise the number of direct discharges. The Environmental Management Act delegates the responsibility of ensuring efficient collection and transport of wastewater to the Municipality. The municipality is obligated to draft an annual environmental and a sewerage plan. This draft has to be publicly available and the Water Boards are able to change it by putting in a petition. According to the Environmental Management Act houses within 40 meters from a sewer should be connected to it. In case of hard-to-reach individual houses subsidies are possible to install septic tanks. The Water Boards can provide permits in such cases. A permit is always required when wastewater is discharged to the surface water directly. The Environmental Management Act describes sewerage performance indicators, known as the "basic effort" (basis inspanning). They point out the required sewerage and pumping capacity in relation to the surface area. Another arrangement between the water boards and the Municipality is the connection permit in which requirements from the Municipality and the water board are described. The Municipalities Act frames regulation related to the taxes a Municipality may charge for providing the sewerage service. Sewerage taxes are not imposed in every Dutch Municipality. Some municipalities finance the sewerage from other public resources. A sewerage benchmark has been introduced as an instrument to improve the efficiency of operation of the sewer systems in various cities/towns of Netherlands. On the initiative of a number of municipalities a foundation RIONED was formed which was entrusted with the responsibility of the benchmark itself (Van den Boogaard and Van

Dijken, 2006). The Pollution of Surface Waters Act came into effect on the 13th of November 1969 and laid down rules on the pollution of surface waters. The act is based on a two pronged approach to fight pollution of surface waters; discharges of polluting substances into surface waters were forbidden without a license and also levies were introduced on discharges according to the polluter pays principle. The polluter pays principle based approach acts as incentive to the polluters to minimise polluting discharges into surface waters as the levies are directly linked to the amount of polluting substances discharged. To set up or operate any establishment a licence from the Ministry of Environment is required. If in addition to this licence a licence for discharge of wastewater is also required, it needs to be applied for, thus helping to identify and charge polluters of surface waters. The industries and the households that cause pollution pay a (pollution) levy to the water boards, which are responsible for treating the wastewaters. The water boards use income from these levies to finance investments required to combat and prevent pollution. Since the year 2000, different water boards have been making comparisons between themselves by using a 'treatment management benchmark'. Comparison of this kind is a suitable instrument for demonstrating the effectiveness and efficiency of wastewater treatment management. Water boards not only use it in an attempt to work with greater transparency in the eyes of taxpayers but also to be able to compare action programmes in order to optimise treatment processes. The water boards charges polluters of the surface and the ground water to protect the raw water sources.

Comparing the Dutch and the UK regulatory regimes with respect to the collection and treatment of wastewater quite similar regulatory instruments are in place. In both cases permits and charges are established to control the sewerage collection and wastewater discharges. Just like for the treating and distribution of drinking water, one of the main differences is that the Dutch situation copes with a fragmented institutional set up. Municipalities are responsible for the sewerage collection, while water boards are in charge of treating the wastewater, which complicates the regulatory interventions. Table 17 provides an overview of the legislation in force for sewerage collection and treatment in England & Wales and the Netherlands.

Table 17 Legislation in force on sewerage collection and treatment

The Netherlands	England & Wales
Environment management Act (1993) Pollution of Surface Waters Act (1969) Disposal Decisions Municipalities Act (1992) Water board Act (1992) North Sea Act	The Water Industry Act (1991) The Water Resources Act (1991) The Water Consolidation (Consequential Provisions) Act (1991) The Environment Act (1995) The Trade Effluents (Prescribed Processes and Substances) Regulations (1989) The Trade Effluents (Prescribed Processes and Substances) (Amendment) Regulations (1990) The Trade Effluents (Prescribed Processes and Substances) Regulations (1992)

Source: Euromarket (2005).

7.3   Regulatory institutions along the strategic components

The regulatory regimes was described in the before section primarily from the perspective of the regula*tor*, i.e. which regulation is in force. Now, the focus will shift towards the regula*ted.*, i.e. how does regulation affect the strategic context of the WSS providers. An analysis is carried out to what extent regulatory institutions constrain or provide opportunities to WSS providers with respect to each of the five strategic components (Boyne and Walker, 2004). Again this analysis will be carried out for the cases of the Netherlands and England & Wales.

7.3.1   Market strategic context

The market strategy addresses the clients the WSS provider wants to serve. Several market strategies are identified which a WSS company may pursue, which will be dealt with in detail below:

1. Market exit.
2. Bulk supply.
3. Common carriage.
4. Inset appointments.
5. Unregulated supply.
6. Mergers and acquisitions within the country.
7. Mergers and acquisitions outside of the country.

Market exit

England and Wales: There is no possibility for the WSS providers for market exit, as they have been licensed in 1989 for a period of at least 25 years for specific geographical regions across England and Wales. However, they do not have exclusive rights and they are not considered as legalised monopolies (Nickson and Muscoe, 2004). The license of the statutory undertaker can be terminated (condition O of the license agreement), although with a ten-year notice. The possibility for market exit is further reduced by the Water Industry Act of 1999 that removed the companies' power to disconnect customers for non-payment of charges. In 2003 over 4.4 million households are in arrears and 3 in 100 never pay at all, causing a bad debt to the water companies of 130 million English pounds annually.

The Netherlands: The Dutch drinking water companies have, just like in England and Wales, the obligation to serve customers located within the assigned region. In the Netherlands this is based upon 30-years concession contracts. Contrary to England and Wales, the Dutch domestic consumers do not have the possibility to select a drinking water company of their choosing. They are 'tied' to their drinking water companies and can be disconnected for non-payment, providing that the company by-laws allow for it. According to data provided by the Dutch Minister of Economic Affairs (Brinkhorst, 2005) an approximate 2.000 out of the 7.3 million connections (0.03%) is temporarily disconnected. These disconnections in general last no longer than several days, and before being disconnected a collection procedure is followed of almost a full year. Within this period defaulters are several times remembered both in writing as by phone, before a debt-collection agency is called in.

Bulk supply

England and Wales. In bulk supply, one water company sells an amount of water to a neighbouring company. Bulk supply is widely practiced in the British water industry (Booker, 1994). As an inter-company trade, these arrangements are negotiated between the two companies themselves and regulation is not interfering, providing that the water resources are not negatively affected and the customers are not negatively implicated. Only in case of dispute between the two companies the regulator (OfWat) can intervene and determine the tariff for bulk water (Jeffery, 1990).

The Netherlands: The Dutch drinking water companies also sell bulk water to each other. This is done through inter-company arrangements that are not regulated.

Common carriage

England and Wales: Common carriage, when one company supplies WSS services to its customers by using another company's network, may include the shared use of a pipe network, treatment works or storage capacity. Common carriage is based on the insistence of OFWAT that operators should grant access to their facilities under certain terms. The guiding principle in England and Wales is that each individual customer is entitled to receive water for domestic purposes from any water company, irrespective of where they live. Although common carriage is conceptually and theoretically interesting, in practice such sharing of facilities runs into severe quality and hygienic complications and has not occurred yet. It is estimated that the potential future market for common carriage is to be around 250 million pounds, relatively small compared to the regulated revenues over 7 billion pounds of the existing water companies (Nickson and Muscoe, 2004). Still, if water companies refuse entrants access to their facilities without an objective justification or on unreasonable terms, they risk infringement of the Competition Act of 1998.

The Netherlands: Dutch domestic customers do not have the right to receive drinking water services from a water company of their choosing. They are restricted to use their current monopoly service provider; hence there is no possibility for common carriage arrangements.

Inset appointments

England and Wales: Another possibility for a water company to enter the market of another English or Welsh water company is by using the mechanism of inset appointments, as allowed and promoted by the regulator OfWat based on the Competition and Services (Utilities) Act of 1992. In this case an interested water company could apply to OfWat to serve (groups of) clients that lie within another operators' supply area, without using the infrastructure of the incumbent operator. These inset appointments are limited to large clients that use more than 100 mega litres of water per year or to sites that are not yet served (green fields). Although the possibility of water companies to make use of inset appointments is available for almost more than 10 years, it has been relatively unsuccessful. Only 9 insets have been approved to date (Nickson and Muscoe, 2004).

The Netherlands: Dutch drinking water companies are allowed to use a mechanism similar to inset appointments. They can compete for so-called 'footloose' customers, being large-scale customers that use water as a means of production. In 2000 the Dutch cabinet decided by instating the Water Supply Act to protect public drinking water companies by forbidding privatisation as far as 'tied' customers are concerned, being households and small industries. But the Water Supply Act opens the market for all other uses, being the 'footloose' customers. Just like in England and Wales, companies that consume more than 100 mega litres of water per year are allowed to choose from which company they buy their water (Kuks, 2003).

Unregulated supply

England and Wales: Another possibility for a drinking water provider to get a new customer in England and Wales is by starting to supply one of the 300,000 customers that currently rely on unregulated supply. Unregulated supply refers to the 50,000 private very small water providers that supply water for domestic purposes to that group. Most of these providers (75%) are single dwellings (Memon and Butler, 2003b).

The Netherlands: The number of potential customers relying on unregulated supply is far less in the Netherlands compared to the UK, as the vast majority of the consumers is already connected. However, water companies can connect those that are not already connected by another water company. Just like the large consumers discussed under the inset appointments these unregulated users are considered 'footloose customers'.

Mergers and acquisitions within the country

England and Wales: In theory, each of the 24 water companies can be bought and sold like any other company (Nickson and Muscoe, 2004), including the possibility of hostile takeovers. Proposals for change of ownership have to be referred to OfWat as the Competition and Services Act clearly states provisions for replacing an appointed undertaker. A recent ruling of the court in the case of Welsh Water insisted that a change of ownership of an established water company should go by a system of competitive bids (Pielen *et al.*, 2004). OfWat is not supportive towards more mergers, as a further reduction in the number of companies affects its ability to make comparisons between companies (Byatt, 1993; Carney, 1990). The Water Industry Act of 1991 requires that the Competition Commission (CC) will be asked for approval if the gross assets of each of the water companies to be merged exceed an amount of 44 million euro. The Competition Act of 1998 outlaws any agreement that (may) have a damaging effect on competition. The Act prohibits agreements between water companies that intend or actively prevent, restrict or distort competition, and also forbids conduct that amounts to the abuse of a dominant position in a market that may affect competition. Hostile takeovers may be allowed. For example, the French company Lyonnaise des Eaux launched a hostile takeover of the licensed operator Northumbrian, which was allowed by the Competition commission. Examples of less successful hostile takeovers are from Severn Trent and Wessex Water, both requesting the CC to takeover South West Water. Both their bids were blocked on the grounds of the loss of information for the regulator and its ability to undertake comparative competition. European restrictions on merging are in place in case the combined aggregate turnover of all the undertakings concerned is more than 5 billion euro.

The Netherlands: the Dutch water companies have the ability to merge with or acquire other Dutch water companies, often this is even promoted by the government. The Dutch water companies have made extensive use of this strategic action. In 1980 there were about 100 companies of which currently just 16 are left (VEWIN, 2003). Supervisory bodies regarded favourably the mergers as they considered a minimum size requirement of 100,000 connections for supply companies to achieve economies of scale (Kuks, 2006). Regulation permits hostile takeovers in the Netherlands. However, few examples are known. In 2000 Nuon Water attempted a hostile takeover of Waterbedrijf Gelderland, but it did not materialize. Just recently, in 2006, the water provider Evides threatened to approach directly the shareholders of the neighbouring water provider Hydron Zuid Holland making them an offer for the shares they owned of Hydron. The management of Hydron Zuid-Holland felt unduly pressurized and warned its shareholders of this possibility. Subsequently, negotiations were obstructed and both parties limited themselves to merely listing the pros and cons of merging.

Mergers and acquisitions outside of the country

England and Wales: With certain exceptions, the regulatory provisions do not apply to business activities of the water companies that are not connected with carrying out their water services in the assigned service area (Jeffery, 1990). To protect the water customers in the assigned monopoly area from losses that could be incurred by other companies within the group, the regulator OfWat ensures that there is no cross-subsidy between the water provider and associated companies (Byatt, 1993). Consequently, the basic organisation of each company is shaped as in illustrated in Figure 16 (Carney, 1990).

Figure 16 Basic organisation of each water company in England and Wales

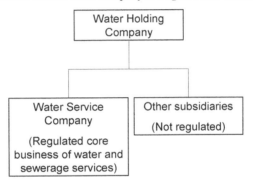

In practice, the licensed water company is often part of a relatively complicated institutional environment. For example, Anglian Water has received a license to provide water services to East Anglia and the East Midlands. For this activity it makes an annual turnover of approximate £900 million with 3,600 employees. Parent company of Anglian Water is the Anglian Water Group (AWG), which also includes the support services group Morrison Plc and AWG Property. AWG has 9,000 staff based in offices across the UK. The group's turnover was £1.5 billion. Again to complicate the corporate structure further, AWG is owned by a private consortium Osprey, comprising of Canada Pension Plan Investment Board, Colonial First State Global Asset Management, Industry Funds Management and 3i Group plc.

The Netherlands: The regulatory environment in the Netherlands discourages water companies to engage into adventures outside the Netherlands. Company by-laws often do not permit operators to expand their activities beyond the service area, they use the argument that the revenues received from the local Dutch customers, should only be used for these local Dutch customers, and not be spent on possibly risky business opportunities outside of the country (Blokland *et al.*, 1999).

7.3.2   Products/services strategic context

Which products/services does the company want to offer? This strategic action concerns the impositions regulatory regimes put on the core business and on non-core activities to the client group that is already served. These two activities are reviewed in the below presented sections.

Core-business of water

England and Wales: According to the Water Industry Act of 1991, water companies cannot escape their statutory duty to deliver "wholesome" water to their assigned monopoly area. "Wholesome" is defined by reference to microbiological and chemical standards and other requirements set out in the Water Supply Regulations of 1989 and 2000. Section 70 of the Water Industry Act specifically makes it a criminal offence for a water company to supply water that is unsuited for human consumption. If a water operator would try to escape their statutory duties, the economic regulator OfWat has the ability to issue an Enforcement Order. If a company does not comply with the order, OfWat can ask the High Court to appoint a special administrator to run the company until arrangements can be made for a new company to take over.

The Netherlands: Just like the English and Welsh companies, the Dutch water companies are forced to provide "wholesome" water. The Ministry of Housing, Spatial Planning and Environment supervises the quality of the drinking water produced. Recently the inspections have been replaced by trust complemented by incidental inspection (Van Dijk *et al.*, 2004).

Non-core activities

England and Wales: Many water companies have diversified into non-water activities. Just like for adventures abroad, these non-core activities are excluded from the regulatory regimes and cross-subsidization is prohibited. In view of the popularity to diversify, regulators are facing a dilemma since diversification of services has meant that companies are increasingly using their core skills and management time on diversification activities. This cannot always be expressed in a price and cross subsidization mechanisms (Booker, 1994).

The Netherlands. The strategic choice to supply non-water services is constrained by regulatory impositions. The Dutch regulators are pursuing a policy of bringing all the activities of public sector organisations that may find themselves competing with market players or market activities into legally financially and organisationally independent units. This means that there is little choice for traditional water companies to transfer their activities other than water supply to a separate company. By dividing their assets in this way they can prevent the business risks associated with their commercial activities being passed on to their tied customers (Kuks, 2006). Company by-laws also often do not permit operators to expand their activities beyond the core business of water (Blokland *et al.*, 1999). On the other hand, the Dutch government has adopted as official policy to pursue cooperation between water managers, municipalities and drinking water companies with a view of making use of environmental opportunities and increasing efficiency (V&W, 1998). Experiments are launched with water chain companies set up to bring together the expertise needed to serve industrial companies and are geared to managing the complete company chain. However, they do not have a multi utility character (Kuks, 2006).

7.3.3   Seeking revenues strategic context

England and Wales: the economic regulator OfWat strictly regulates the price setting of the drinking water since the Competition and Service Act was passed in 1992. OfWat sets the allowable price (also known as the K-factor) based upon the price-cap mechanism that the companies are obliged to follow. Every five years K is calculated, taking account general retail price inflation as well as results from the yardstick competition that OfWat is conducting. Since K is applied to a basket of regulated charges, covering both measured and unmeasured water and sewerage charges, as well as trade effluent charges; each company still has the ability to increase or decrease average charges for individual "basket items".

The Netherlands: In the Netherlands the tariff is set by the company itself. The management of the water company prepares an annual proposal for the tariff level and structure for approval to the company shareholders based on the notion of cost recovery. Municipal and provincial governments as owners of the water providers, have the power than to agree with or reject the proposed tariff system. National government does not interfere in matters of water pricing. Hence, the only regulation enforced upon the tariff is the control of the shareholders to which the tariff needs to be proposed, although even that depends on the company bylaws. Since these shareholders are public entities such as municipalities or provinces, the Dutch regulatory system assumes this is sufficient to maintain a level of equity and affordability.

7.3.4 Internal organisation strategic context

How to structure the internal organisation? This strategic action focuses on internal organizational arrangements for service provision.

England and Wales: The English and Welsh companies have complete management control, although they need to consider specific regulatory provisions as for example the Guaranteed Standard Scheme (GSS), which contains information on customers' rights (including compensation for supply interruptions). Yardstick competition is not only useful in respect to determining the price cap but is intended also to work as a motivation to improve performance. Participation in the yardstick comparison is compulsory (Robinson, 1997).

The Netherlands: Just as the English and Welsh water companies, the Dutch public limited companies have complete managerial autonomy, although the company bylaws limit the manager in its freedom of operations. For the organisation of the drinking water supply service, most of the Dutch drinking water companies have embraced a system of voluntary benchmarking since 1997, looking at four aspects: drinking water supply, cost efficiency, environmental performance and service performance. The objective of this benchmark is to increase the transparency of the drinking water companies' performance and to provide an instrument that can improve efficiency.

7.3.5 External organisation strategic context

How to interact with external parties? External organisation refers to the inter-organisational relationships through which many organisations provide services. A distinction can be made between:
1. The relation with the "supplier" of raw water.
2. The relation with the suppliers of subcontracts and other materials.

The relation with the "supplier" of raw water
England and Wales: the Environmental Agency, mandated by the Environmental Act of 1995, monitors continuously the amount of environmental pollution generated by

the water companies including their effect on the water resources due to abstraction. To abstract water, companies need to apply for a time limited abstraction licence issued by the Environmental Agency. In case a licensed abstraction causes damage or loss to anyone, the person has the right to seek financial compensation from the abstractor. Moreover, water companies are obliged to submit water resource plans each year to the Environmental Agency.

The Netherlands: The water company is dependent on receiving a license to extract groundwater. The Province issues these licenses. Moreover water companies need to pay an abstraction charge of about 34 cents per cubic meter. As the regulatory bodies want to discourage the use of groundwater, there is rebate of 28.5 cents per cubic meter in case surface water is injected into the groundwater prior to injection (Van Dijk *et al.*, 2004).

The relation with the suppliers of subcontracts and other materials.
England and Wales: For engaging external parties, the Utility Contract Regulations dating from 1996 regulate the procurement in the water sector and other utility industries. On the basis of this regulation OfWat monitors companies' use of associates for subcontracting.

The Netherlands: Just like in England and Wales, Dutch water companies have to comply with procurement rules for tendering and bidding as formulated by the European Commission.

7.4   Synthesis of the Chapter

Individual countries impose regulatory regimes for WSS providers. The analysis of the current situation of the Dutch and the English and Welsh WSS sectors through the elements of the water cycle provides an insight in the segregation the Dutch policy makers made in resource access, water treatment, water distribution, sewerage and wastewater treatment. From the presentation of the regulatory objectives and instruments it can be concluded that the main difference between the regulatory regimes is that the regulatory responsibilities are much more scattered over various entities in the Netherlands compared to England & Wales. However, investigating the content of the legislation only small differences can be observed for all elements of the WSS cycle.

With respect to the possible influence different institutions may exercise on the strategic context (managerial discretion) of the WSS providers a subdivision is made along the different strategic components (market, products/services, seeking revenues, internal and external organisation). A different institutional context may be conducive to formulating different strategies. From the analysis we conclude that, at least conceptually, the regulatory institutions in the two countries differ. In England & Wales, all customers are legally entitled, although difficult in practice, to receive water from any water company, irrespective of where they are located. The Dutch approach is more restrictive. They classify customers into 'tied' and 'footloose' customers, whereby only the footloose customers have such entitlement. Consequently, the English and Welsh regulatory context allows water companies to chase for customers while such possibilities are more limited in the Netherlands. This may again be the basis why the regulator in England & Wales is strongly constraining the possibilities for a water company to merge within another English or Welsh water company eyeing to preserve both the possibility for competition in the market, and comparative competition. On the other hand, the regulator is lenient for English and Welsh companies if the ownership goes from one private owner to another, or even if the water company starts to buy or merge with water companies outside of England and Wales. In the Dutch case it seems to be just the other way around. The government has actively promoted mergers within the Netherlands during the last decades, but shifts in ownership are tightly regulated. The Dutch water companies are very much discouraged, or even prevented to merge with or to be taken over by a foreign company.

Table 18 Summarized opportunities and constraints from regulatory regimes

Regulatory regime on		Private water companies (England and Wales)		Public water companies (The Netherlands)	
		Regulatory Opportunities	Regulatory Constraints	Regulatory Opportunities	Regulatory Constraints
Strategic Actions	Market	*No exclusive rights given to incumbents *To use bulk supply, common carriage & inset appointments *Transfer of ownership *Activities outside of E&W	*Bound to assigned population. *Cannot disconnect customers *No mergers inside E&W	*To use bulk supply *To benefit from inset appointments * Mergers inside the Netherlands allowed	*Bound to assigned population *Exclusive rights of incumbents *No change ownership *Activities outside the Netherlands not encouraged
	Services	*Differentiation	*Bound to supply wholesome water *No cross-subsidisation allowed		*Bound to supply wholesome water *Separate other from water activities *Often restricted by by-laws
	Revenues	*Appeal to tariff setting by OFWAT *Indirect influence through negotiations and participating in yardstick	*Price cap regime. *Limitations to compulsory metering.	*Company sets itself the tariff individually *No limits to compulsory metering	*Approval of shareholders
	Internal organization	*Each company has complete management control	*Acknowledge customer rights. *Compulsory yardstick participation.	*Each company has complete management control *Voluntary benchmarking	
	External organization		*Abstraction licenses *Submit water resource plans *Procurement rules for tendering		*Abstraction licenses *Procurement rules for tendering

Source: Schouten and Van Dijk (forthcoming).

The same conservative attitude of the Dutch regulatory regime is applicable for the provision of non-water services. Often the company by-laws prevent such undertakings. The English and Welsh regulator is staying out of the diversification decision, although the regulator prohibits any cross-subsidisation from its water business. For short, mergers and acquisitions in the Netherlands are limited to mergers between Dutch public owned water companies, while in England and Wales water companies are not allowed to merge, except with foreign companies.

The larger discretion English and Welsh water companies enjoy with respect to market and product strategies may be a consequence of their smaller discretion with respect to tariff setting. Tariffs are strictly regulated in England & Wales, while in the Netherlands tariff setting is left largely to the discretion of company management.

In sum, this Chapter provides a first step in answering the question to what extent WSS operators operating under a different institutional arrangement pursue different strategies compared to publicly owned water companies. The main conclusion from this Chapter is that regulatory institutions in England and Wales and the Netherlands invite different strategic directions. In Table 18 all the regulatory constraints and opportunities for the five strategic actions are summarized and compared.

Section III. Analysis

Chapter 8 Institutions and the strategic plans of WSS providers

This Chapter is concerned with the second level of analysis of the neo-liberal implications on the WSS sector. It aims to identify whether neo-liberal reforms measures may have an effect on the strategic plans of WSS providers. In this pursuit, a case study is conducted in the Netherlands Antilles.[50]

8.1   Introduction

This Chapter aims to bring further insight to how the strategic plans of WSS operators in different institutional contexts compare. The main part of the Chapter comprises a case study located on the Caribbean Island of St. Maarten. In the case, one public and one private party compete to be given the opportunity to manage and extend the WSS services at the Island for the coming 10 years. The author was involved as an independent evaluator, and received permission from the government to use the case for scholarly purposes. This benefited the research greatly in terms of access to full information and gaining inside knowledge. The collected information is formatted in this case study to assess in a structured manner the differences between both parties' intended strategies to manage the service provision.

To test whether private WSS providers have different strategic plans compared to public WSS providers, the bidding documents are analysed for a specific tendering event, e.g. a 10 year contract in St. Maarten. In the following section the case study is elaborated. The case study description is structured in the following manner. First the background is given of the ongoing trend to involve private parties in the Caribbean WSS sector. This is followed by a more detailed analysis of the specific context and developments at the Island of St. Maarten. The decision process leading towards (possible) private sector involvement is reviewed in detail. Then, the actual comparison is presented between the strategic plans of both parties, along the dimensions of Boyne and Walker (2004). The case concludes by making some final comments.

8.2   Case study St. Maarten

St. Maarten is one of the five islands that form the Netherlands Antilles, which in turn is a part of the Kingdom of the Netherlands. The local government of St. Maarten has the authority to handle the most vital functions; however, they cannot conflict with the laws of the Dutch Antilles or the Netherlands. The island is located approximately 280 miles east of Puerto Rico (see Figure 17).

---

[50] Parts from this Chapter have been separately published as: Schouten, M., Brdjanovic, D. and M.P. van Dijk (2008). *A Caribbean Evaluation of Public versus Private Drinking Water Provision: the case of St. Maarten, Netherlands Antilles.* In: International Journal for Water. Volume 4, No. 3/4; pp. 258-274. Inderscience Enterprises. Ltd.

Figure 17 Map of St. Maarten/St. Martin

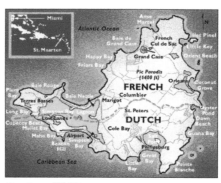

Source: Sint Maarten Tourist Brochure (2007).

St. Maarten/St. Martin is a blend of two European cultures, namely the French and the Dutch, set in the Caribbean. The island of St Maarten has an interesting historical background; Christopher Columbus discovered St. Maarten in 1493, on the Patron Saint Day of St. Martin of Tours, and in the 140 years that followed, it changed flags many times. The Spanish, French, Dutch and English all claimed possession in 1648, when the island was divided in two by the Treaty of Concordia, which was signed on 23rd March 1648. The Dutch received 16 square miles and the French received 21. St Maarten, with its capital Philipsburg, occupies the southern part of the island. Today, the inhabitants of this friendly island are proud of their 350-year coexistence and have never constructed an official border, having only a symbolic border between the island's two countries. There is no border control between the two parts of the island; there is free access and movement between the two countries and border signs indicate the entrance from one side to the other. The French part of the Island, St. Martin, is out of the scope of the case study as it separately governed and its WSS services are also separately organised.

St. Maarten has enjoyed over the last few decades a tremendous economic growth. Annually almost 1.3 million tourists visit the island. This total includes both tourists on cruise ships and those who actually sleep on the island. The boom in tourism on the Island in the 1990's resulted in an explosive growth in population. At the last census count in 2000, the population of Dutch part St. Maarten was around 43,000 people, made up of almost 100 nationalities, very much in contrast with the situation in 1960 when it counted only 3,000 inhabitants. The explosive increase of tourism and its accompanying economic growth has resulted in an increased demand for larger coverage and better quality of utility services (Brdjanovic and Gijzen, 2005). However, most infrastructures required to deal with the consequences of economic development are lagging behind. Especially sewerage networks are insufficiently developed as large parts of the population are not connected, and wastewater is often inadequately treated prior to discharge, which can cause health problems as well as environmental damage. Owing to the relatively hilly area, the distribution system is hydraulically complex, although relatively small. There are seasonal peaks in demand to be coped with coinciding with the peak in hotel occupancy at the end of the dry season (World Bank, 2005a). Due to its' small size, lack of natural water storage, and its' vulnerability to natural and anthropogenic hazards including drought, cyclones and urban pollution, St. Maarten is coping with fragile and scarce water resources and insufficient water infrastructure (World Bank, 2005b). Despite its urgency for change a common denominator with respect to service provision (especially wastewater services) is the local population's insufficient awareness of the need for environmental protection and conservation and this lack of awareness is one of the main reasons that governments unfortunately often do not consider remedial actions as a priority.

Hence, the Island government is continuously challenged to come up with, and apply, both innovative solutions for funding and managerial arrangements in order to provide and maintain sustainable WSS provision. Although the Island government of St. Maarten has acknowledged the need for mitigating measures in establishing a proper infrastructure to keep up with the rapid growth of the population, it has been rather slow in realising this. The WSS services are put under the responsibility of different parties in the following manner, obstructing to attain the benefits from an integrated water cycle approach.

1. *Drinking water treatment.* The Island relies for its drinking water treatment on a reverse osmosis membrane plant with a maximum production capacity of 13,500 cubic meter per day, which desalinates seawater. Although this plant is owned by the government, a 10-year operation and management contract was signed in 1997 with a subsidiary of the French multinational Veolia.
2. *Drinking water distribution.* Drinking water distribution has been the responsibility since 1970 of a publicly owned multi-utility, which has been responsible for both electricity and drinking water distribution for the Dutch part of Sint Maarten. The company procures the treated water from the operator of the desalination plant, distributes it, and collects the water tariff from the customers.
3. *Sewage collection.* Although the majority of St. Maarten is unsewered, there is a limited sewage system at the Island. The current number of house connections is estimated to be around 1,000, mainly within the areas of the capital Philipsburg and a few other locations. The government of St. Maarten owns the sewage facilities, but the sewage collection operations are outsourced via a management contract to a local private contractor.
4. *Storm water collection.* In the tropical climate of St. Maarten, heavy rains and storms are common. Sometimes hourly and daily precipitation at some locations can be as much as 150 and 600 mm respectively! Due to the small size of the island and its steep slopes, it does not take long until this mixture reaches beaches facing the open sea, seawater lagoons, fresh water ponds or brackish inland water ponds. It is currently impossible to estimate where the environmental effect is greatest, especially in relation to the water quality. It could be on the bathing waters of the picturesque beaches, on the surrounding coral reefs that provide a habitat for marine life (including the legally protected turtles), on the fragile ecosystem which hosts a number of rare bird species, or maybe on the already heavily polluted waters of one of the ponds of the Island.
5. *Sewage treatment.* The total capacity of the wastewater treatment facilities at St. Maarten have a design capacity for 5,000 to 6,000 users, covering an approximate 15% of the total demand. Besides these, there are several private package plants owned by hotels. The majority of plants are activated sludge plants, and some employ attached-growth systems (such as Rotating Biological Contactors). In general all the plants perform poorly (Brdjanovic *et al.*, 2005). According to current practice, limited attention is paid to sludge treatment; the largest treatment plant has some provision for sludge drying, while the other plants have no sludge-

related facilities whatsoever. The package plants associated with the hotels are not under the jurisdiction of the Island Government and are operated and controlled by the hotels themselves. Laboratory facilities are limited and poorly equipped. Furthermore, there are no wastewater effluent standards or regulations in place, nor is there any sort of continuous monitoring of the surface water quality (Grabowski & Poort, 1995; 1998).

After the destruction in 2000 caused by Hurricane Louis, the existing water infrastructure needed to be urgently rehabilitated. The first action taken by the government, being part of the Kingdom of the Netherlands was to seek financial support from the Dutch government to reconstruct the destroyed sewerage system. Although their request was granted, the Dutch funds could only be disbursed if the Island fulfilled several conditions. To satisfy these conditions, the Island government hired several consultants to reassess the scope of the project. Based on that assessment the Island government became convinced that to thoroughly tackle the water and wastewater problems of the Island, it should take the matter at hand much more rigorously. Possibly inspired by the popularity of private sector involvement in neighbouring Caribbean Island states (see Box 3), the St. Maarten government started to explore the possibility to substitute the traditional local providers of WSS services by private companies, hoping such would bring an additional impetus to the service provision.

Box 3 Popularity of private sector involvement in the Caribbean

Private sector involvement as an innovative way to attain sustainable water WSS provision has been popular amongst Caribbean governments. Several examples can be identified of private sector involvement in the WSS sector (Pinsent Masons, 2006); of which probably the most famous one is the contract in Puerto Rico due to the extensive media coverage. In 1995, the Puerto Rico Aqueduct and Sewer Authority (PRASA) signed a management contract with Veolia, which lasted seven years. After a critical assessment of its performance, Veolia was replaced by another private operator (Suez). Suez signed a 10 year operations and maintenance contract, but also this contract was prematurely terminated in 2004. Currently the public authorities of Puerto Rico are resuming the responsibility for the management of WSS services on the island. Another well known example is the 50/50 joint venture between Aguas de Barcelona and the Cuban government, dating from 1999. The joint venture is serving a total of 1.8 million people in Havana with a duration of 25 years. Also in Trinidad and Tobago a joint venture was established between Severn Trent, WASA and the Government of Trinidad and Tobago. This contract was supported by a US$ 80 million loan from the World Bank, and ended in April 1999.

The Dominican Republic signed in 2001, two service contracts with private parties for the installation of meters, meter reading, billing and collection, one for the Eastern and one for the Western part of its service area in Santo Domingo. The contract for the Western part was awarded to the Colombian company AAA, which increased the share of metering from 1% to 25% and increased collected revenues by 128% in less than two years. Examples of Build Operate Transfer (BOT) type of contracts can be found on the British Virgin Islands, the Cayman Islands and the Bahamas.

The major reason for the relatively high degree of private sector involvement in the Caribbean region is that a lot of Caribbean islands experience difficulties in keeping pace with the claims that economic development places on the infrastructure of the islands (UNEP, 1999). The Caribbean has succeeded over the past decades in realizing a sustained growth in per capita incomes, with most of them becoming middle-income countries and achieving high levels of economic development. The average growth in GDP per capita over the last 40 years for a median Caribbean country is 2.8% higher than that for Latin America (World Bank, 2005b). Yet, currently the abiding impression is one of concerns for the sustainability of the past accomplishments. The most important asset that brought about the economic development, being the beauty of the natural environment, is severely under threat. The quotation 'Beauty is a fragile gift' from the Roman poet Publius Ovidius Naso (43 B.C.) more than two thousand years ago, seems in this respect all the more relevant to the Caribbean islands in the present day. Another reason for the high degree of private sector involvement is the attractiveness of the region for private parties. The region contains many densely, middle income and rapidly growing urbanised areas. The attractiveness for private parties to get involved into the service provision in the Caribbean is strengthened by the favourable conditions imposed by international financial institutions with loan (re-)negotiations and structural adjustment programmes. A final reason why the Caribbean Islands have a high degree of private sector involvement relates to the scarcity at the Island states of adequate human resources. The small size of the Islands creates a problem for responsible governments and local service providers to access adequate human capacity, skills, and financial resources. External introduction of knowledge and skills by outsiders (foreign private parties) might bring the necessary human and financial resources to the Islands.

Exploring the possibility of private sector involvement in the WSS sector the Island government of St. Maarten got into contact with a private Dutch party that proposed the idea of establishing a joint venture company that would be made responsible to construct and manage a large-scale sewerage system on the Island. In principle, the

Island government positively received this idea although it was hesitant on the distribution of shares in the to-be-established company. Initially it was the intention that the Island would only have a minority of the shares. For the private party this affected the feasibility of the project and as compensation it proposed to include also the drinking water provision into the scope of the to-be-established company. In this way important synergies could arise that would make the project feasible.

The inclusion of the drinking water provision in the scope of the project proved to be an important landmark in the decision process. A major actor in the St. Maarten community, the monopolistic electricity and drinking water provider, became involved. This public company and its employees, who were united in a trade union, did not care much for splitting off the drinking water activities to a foreign Dutch company and requested the opportunity to make a counter offer. The Island government again was open to this suggestion and asked both parties, the Dutch private party and the local electricity and drinking provider, to make a final bid. The government made the public interest at stake explicit by defining what it expected from the winning party (see Table 19).

Table 19 Explicit expectations of the Island government

- To connect as many households and businesses to the water network as is financially feasible.
- To meet the international standards in terms of the quality of the distribution network, whereby outdated lines will be replaced by newer lines of sufficient capacity and the losses due to leakage will be reduced to acceptable levels.
- To cover the costs associated with the laying and maintenance of the distribution system with the current drinking water tariffs
- To achieve insofar as is financially feasible, the collection, transportation and treatment of sewage for the Dutch side of St. Maarten.
- To improve the quality of the surface water to acceptable levels.
- To achieve the highest possible protection of the environment.
- To minimize the degree of problems or hindrance that is caused by untreated water after heavy rainfall.
- To manage the sewage and treatment facilities in an adequate fashion.
- To recover the costs of laying and maintaining the sewage and treatment facilities from the polluter by means of an acceptable tariff on environmental and sewage privileges.

Although the foreign private party was not amused by the inclusion of another party in the tendering process, both parties honoured the request and on 31st May 2002 new proposals were submitted. It took until October 2004 to approve the funding and the appointment of the independent evaluator to conduct the evaluation of the two proposals. Based on the independent evaluation of the two bids the Island government would make its decision. Once the evaluation was concluded in 2005 the conclusion was clear. Both parties did not manage to put a convincing proposal on the table although in comparison the proposal of the local public electricity and drinking water provider was the least weak one. The local company's proposal scored better on seven criteria, while the private party outscored the public party on only one criterion, and on two criteria, they approximately drew. It is interesting to notice that the theoretical maximal score (in the case of a perfect proposal) in the Multi Criteria Analysis (MCA) would be 10,000 points. The 'winning' party scored only approximately 35% of the maximum possible score, and its winning margin over the private party's proposal was a rather modest 8% of the achieved score. Hence, an important confirmation provided by the MCA is that neither proposal achieved more than approximately one third of the possible total score. Neither of the proposals fully complied with the given Terms of Reference nor is of the professional level expected for proposals for a project of this size, importance and potential impact. Based on that conclusion, the Island government has entered –and at the moment of writing still is– negotiations with the local public electricity and drinking water provider how to best include the responsibility for extending and managing the WSS services to the St. Maarten population.

The bidding documents submitted by both the public party and the private party for a tendering organised by the government of St. Maarten provide the basis for comparing how both a public and a private party compare with respect to their intended strategies for service provision. Although the conclusion of the appointed evaluators was that the public party was the better one, their analysis does not surface well how both parties compared in their intentions towards service provision. The evaluation was carried out on the basis of a predetermined set of ten evaluation criteria, each having a different weight (Brdjanovic *et al.*, 2005). As the evaluation criteria, which were predefined by the Island government, were conceptualised on the basis of the policy priorities of the Island, they can only partly support a comparison of the intentions of both bidders in terms of conduct. To structure the comparison of the intended strategies use is made of the distinction Boyne and Walker (2004) into five strategic components that public service organisations may have, i.e.:

1.   The clients the party wants to serve.
2.   The products or services the party wants to offer.
3.   The management of costs and revenues of the operations.
4.   The internal organisation of the operation of services.
5.   The external relations the party wants to engage in.

The following Table 20 divides the evaluation criteria, as set by the Island Government along the above five strategic components of Boyne and Walker.

Table 20 Evaluation criteria categorized along the strategic components

Strategic action	Tendering evaluation criteria
Market	• The degree of coverage of water and sewer connections.
Products/Services	• The WSS services delivered. • The amount of water distributed (water treated minus Unaccounted for Water)
Seeking revenues	• The investment costs. • The financing arrangements for investment, including donor financing. • The return on investment and solvency. • The charges to the customers.
Internal Organization	• The degree in which use is made of local resources. • The levels of synergy. • The degree to which consideration is given to the interests of personnel currently involved with sewage collection and treatment, and drinking water distribution facilities. • The legal organization for the execution of the project as well as the responsibility for, and authority and control over, the execution.
External Organization	• The conditions that are being demanded of the Island government to guarantee the successful execution of the project.

In the analysis for each strategic component, first the expectations and constraints given by the government (based on the tendering document) will be described, before the intentions of both parties are compared. These governmental requirements might limit the discretion of the bidding parties in their intentions. Hence, the first issue to be identified in the analysis of each strategic component is the extent to which the bidding party is limiting itself by responding strictly to governmental references, or whether it allows itself to pursue ideas of its own. The second issue to be compared is the extent to which one bidder's intentions can be identified as conservative or opportunistic. Does the party intend to make use of innovative, risky strategies to achieve its objectives, or does it rely on established technology and practices? The two issues (Reactor-like and Prospector-like) reflect the two continuums forming the basis of the typologies of Miles and Snow (1978) to distinguish strategic typologies. The two questions are related to each other, i.e. a company can be reactive but conservative, or reactive but opportunistic. Also the party can be non-reactive but conservative, and vice versa.

8.2.1   Strategic market plans

The government instructed both parties to connect as many households and businesses to the water and sewerage network as financially feasible. The government did not set a minimum or a maximum number of customers to be connected, and left this to the bidding parties to propose.

Coverage. The differences between the two proposals were very small related to the total numbers of customers they intended to connect to the water and sewerage network. Both parties, used available population growth statistics and reference documents (Grabowski & Poort, 1998), and based on these they identified the number of people to be connected to the water and sewerage networks. The proposed number of connections to the drinking water network after 10 years is almost identical (15,000 for the private party versus 15,060 for the public party). Also the figure for inhabitant's equivalent to be sewered after 10 years was close (65,700 for the private party versus 64,000 for the public party). Hence, although the government left the total number of people to be connected to the water infrastructure relatively open, both parties were similar in their response.

However, a major difference between both parties was the speed in which the new customers were to be connected. The private party aimed to connect and construct as many connections as soon as possible, while the public local party stretches the construction work evenly over the entire 10-year period. A clear motivation for the private party to quickly connect new customers is that such creates an immediate income stream from these new customers, strengthening the financial viability of the proposal. The public party apparently did not perceive such pressure. Hence, the private party can be assessed as more aggressive in its intentions in view of its strategy to connect people quickly to the network.

8.2.2   Strategic products/services plans

The government explicitly asked within the Terms of Reference to meet the international standards in terms of the quality of the infrastructure facilities. The technology to be used to reach these required levels was left to the discretion of both parties to propose. Next, it also requested the bidders to reduce the losses due to leakage to acceptable levels.

The drinking water and wastewater services delivered. The differences between both parties were marginal. For drinking water, both parties relied on the same desalination plant to produce drinking water. Related to wastewater services provision, both parties based themselves on consultancy reports that were years before submitted to the Island government. Within these reports several recommendations were included which were all adopted by both the bidders. Both parties planned to use the same technology for wastewater collection (separate gravity sewers) and for wastewater treatment (an activated sludge Carousel©). Hence, although there was sufficient discretion for both parties to be innovative in their intentions, neither of them took this freedom but just relied on earlier made recommendations, possibly tempted by the convenience to save time by using the existing literature.

The amount of water distributed (water treated minus Unaccounted for Water). Currently the level of Unaccounted for Water (UfW) is approximately 25%. Since the treated water is procured from the operator of the desalination plant, the leakages reflect a financial burden and importance is given to reduce the leakage in the distribution network. In this respect, the private party sets a more ambitious goal towards reducing the level of Unaccounted for Water (10% UfW compared to 15% UfW) and can, hence, be identified as less conservative compared to the public party.

8.2.3   Strategic seeking revenues plans

The government did not set any requirements on the amount of planned investments of the bidders, although it did set some references with respect to the financing of the operations. The government preset in the tendering documents the level and structure of the water and sewerage tariff for the coming 10 years. Hence, both bidders had no possibility to shape the tariff differently.

The investments. The private party proposed by far the largest investment (US$ 122 million) over the coming 10 years, almost 70% more than the public party (US$ 72 million). The reasons why the private party required much more investments can be derived from Table 21.

Table 21 Overview of investment by the public and the private party

Investment element	Private party US$	Public party US$
1. Takeover of existing water infrastructure	17 million	0
2. Takeover of existing desalination plant	11 million	0
3. Additional wastewater facilities	74 million	62 million
4. Additional drinking water facilities	20 million	10 million
5. Institutional strengthening	0	300 thousand
TOTAL	**122 million**	**72 million**

The Table shows that although both for drinking water as for wastewater the direct investment of the private party are more compared to the public party's intentions (an additional US$ 22 million investment), a substantial part of the high investment part of the private company is explained by the costs associated with taking over the current infrastructure. Since the private party intended to set up a joint venture company that would own the infrastructure it managed, it had to buy the existing infrastructure from the current owners, in contrast to the public party. Also the phasing of the investment was very different the private party proposed to put the focal point of the investment at the beginning of the ten-year period, to boost the number of connections as soon as possible, while the public party proposed to spread it more evenly over the same period. Another argument for the private party's proportionally high investments at the early stages of the contract might be the reluctance to invest close to the expiry date of the contract, as it will not be able to reap the fruits from these late investments.

Graph 2 Phasing of planned investments of the private versus the public party

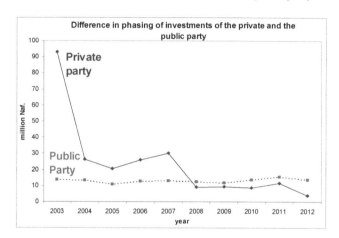

The financing arrangements for investment, including donor financing. Furthermore, the envisaged financing arrangements were conceptually different. The public multi utility would rely mostly on internal cash flow generated from the electricity sales. The foreign private party intended to rely mainly on attracting funds from external parties such as banks and donors. Both parties assumed the availability of the US$ 27 million as a grant from the Dutch Government. An important difference between the two proposals is that in the case of the local multi utility, such a grant was considered as optional, but the private party relied on this grant to keep the tariffs at an acceptable level and to limit the amounts to be borrowed commercially. The different approach of the private party made it more vulnerable compared to the public party in many aspects:

- Apart from the assumption of the private party that it would be able to source the Governmental grant, it also assumed it would be able to find for the remaining investment amount sufficient and cheap external funds on the commercial market.
- The private party depended on the government, the contractor of the water treatment plant and even the public multi utility for the transfer and procurement of the existing assets. If any of these parties were not willing or able to co-operate, it would severely undermine the overall project.
- The private party relied on the assumption that it would be able to generate an income stream from connected consumers relatively soon in the project. If there were delays in the implementation schedule, this would create a problem in the financial viability of the overall project.

- The public multi utility had proven its capability in collecting (water and electricity) fees and its willingness to take measures with outstanding debts during its operations on the island. The private party as a newcomer did not have this local experience.

A common conclusion made for both proposals was that WSS services would not be a money-maker in the near future in view of the large amounts of investment to be made. The private party can be identified as more ambitious and opportunistic, in view of its higher and faster investment programme, and its more risky financing arrangement.

8.2.4   Strategic internal organisation plans

The government instructed the bidders to manage the infrastructure in an adequate fashion. However, to define what 'an adequate fashion' entails was left to the bidders. A specific criterion was to what extent the bidder intended to make use of local resources, how it dealt with the current personnel involved, and how it intended to achieve synergetic benefits.

The degree in which use is made of local resources. Both bidders suggest that use will be made of local resources seeking in this respect compliance to the governmental references; although realistically the degree to which both parties will succeed in involving local resources largely depends on their availability on the Island.

The degree to which consideration is given in the proposal to the interests of personnel currently involved with sewage collection and treatment, and drinking water distribution facilities. Compared to the public company, the private company would have a much larger challenge to address issues related to the personnel currently employed. The possibility of a transfer to the newly to-be-established company was one of the major reasons for resentment among the employees of the public multi utility against the foreign private party, and had contributed to the gradual development of a very vocal opposition against involving a company from the Netherlands. The local company's proposal would cause fewer disturbances to current employment levels as most employees would be able to remain with the same employer. It could be anticipated that if the private party were given the project then numerous obstacles in its implementation concerning labour issues would arise.

The levels of synergy. Both parties stated they wanted to achieve synergistic benefits, although the scope to achieve these synergetic savings was different for both bids. The local multi utility would try to realise them between its electricity operations, its current water distribution operations and the new wastewater operations. The foreign private party would try to establish them between the water treatment, water distribution and the wastewater operations. Not only was the scope different, also the estimated amount and how sure the bidders were in realizing these savings, differed. The local public company estimated total synergetic savings of US$ 17 million while the private company estimated them in a range between US$ 11 to US$ 17 million. The private bidders was less sure of realizing these investment savings, possibly because its dependence on other parties.

The legal organization for the execution of the project as well as the responsibility for, and authority and control over, the execution. The major difference related to the legal organisation between the two proposals was the establishment of a new company versus an existing one. The liability structure of the existing local multi utility was clear; in the private option it still had to be seen. A disadvantage of the local company's proposal was the complete integration of almost all utility services into a single integrated company. This implied an extreme dependence of the government and the whole economy of the Island on one utility concerning pricing and service delivery. Moreover, such integration implicated the continued acceptance of cross-subsidies from electricity towards water and sanitation. The private option was complicated as far as the transfer of assets was concerned. It required agreement on the water distribution part to be transferred to the newly created company. This proposed transfer would encounter some obstacles in implementation. In its accounting reports the local multi utility put a claim on the water distribution assets of about US$ 56 million as previous losses on the water distribution activities. If the Island Government wanted to transfer these assets, it could be expected that the public multi utility would ask for repayment of these losses, slowing down the process substantially. Another difference was that in the local public option, control by the government is more direct, since the government owns the company, nominates its director and a commissioner is chair of the Board. In the case of the foreign private party the government is a minority shareholder and can only indirectly execute authority.

In sum, with respect to how the parties intended to internally organise themselves, both parties can be assessed as having a same level of compliance to the governmental references. The private party's intentions can however be assessed as more opportunistic, especially with respect to its legal organisation, and taking over of personnel.

8.2.5   Strategic external organisation plans

The Island government has to establish a facilitating environment in order to make the project successful. In view of the limited institutional capabilities of the Island government, the smaller the demands of the bidder, the higher the proposal scored.

The conditions that are being demanded of the Island government to guarantee the successful execution of the project. Due to the more intensive investment that the proposal of the private party envisaged at the beginning of the project, it is expected that the private party will need more support from the government in bridging the expected problems regarding disruption of life on the island by construction works. Moreover, the private party identified many conditions for the Island Government, of which several of them seemed very hard to satisfy. Especially in view of the weak financial condition of the Island Government and the latent shortage of qualified civil servants, it can be predicted that these will form large obstacles in an implementation phase. Hence, the public option seemed the most practical choice here, although opting for the public local multi utility in fact implies the creation of a large monopoly on the Island. Hence, again the private party inhibits the most risk taking behaviour compared to the public party.

8.3   Synthesis of the Chapter

To analyse whether WSS providers with different institutional arrangements formulate their strategic plans differently, a single case study in St. Maarten was conducted in this Chapter. To represent the strategic plans of the WSS providers the bidding documents were used, which both a public and a private party have submitted to be granted the opportunity to manage (and extend) the WSS provision at the Island of St. Maarten.

The St. Maarten case is typical for the whole of the Caribbean region where policy makers are exploring alternative institutions to meet the increasing demand for drinking water and to upgrade the currently lagging public infrastructures on the Island. Due to a tremendous economic growth the Islands' population, and the number of tourists visiting the Island, have steadily increased. The Islands' policy makers are continuously challenged to ensure that the water and wastewater infrastructure is able to cope with the demand. After hurricane Louis destroyed much of the infrastructure on the Island in the year 2000, there was an immediate urgency to reconstruct the infrastructure. In finding ways (and finances) to realize such, the Island government started to explore the possibility to involve private parties on a large scale in the WSS service provision. It is noted that in the Caribbean many other governments have preceded St. Maarten in exploring private sector involvement as an alternative to public sector management.

Hence, the Island Government issued a tendering procedure inviting parties to come forward and submit a bid to manage and extend the full WSS provision for the coming 10 years. Two parties reacted. The first one was a foreign Dutch private party who proposed to set up a joint venture in which the government would hold a minority share. The other one interested to win the contract was the incumbent local publicly owned utility in charge of drinking water provision. In the external evaluation carried out it was found that the incumbent public party managed to propose the better proposal. What can be observed from the case study is that the decision process in weighing between the private and public option should be carefully constructed including the involvement of stakeholders, taking the history and the local institutions into account; and even then there is no guarantee for success.

In this Chapter the final result of the external evaluation is not our primary interest. An analysis of the contents of the two bidding documents along the five strategic dimensions (markets, products/services, seeking revenues, external and internal organisation) is more valuable to see whether any differences may be explained by the different institutions of the bidders. The assessment of the intentions of both bidders centred on two main questions, i.e. the extent to which the party took the freedom to deviate from the governmental references, and secondly the extent to which the intentions of the party were risky and innovative.

For all of the five strategic dimensions, the plans of the private party were identified as more aggressive, opportunistic and innovative. Neither of them took the effort to deviate from these references, using their expertise and creativity, adding additional value to their proposals.

- With respect to the strategic market plans both parties relied on governmental references, however the private party was much more aggressive in its connection policy.
- The higher level of ambition of the private party was also recognised for the strategic products/services plans. The private party wanted to make larger reductions in the levels of leakage compared to the public party.
- Regarding the strategic seeking revenues plans both parties had to adhere to the tariffs given by the government. However, large differences were noticed between the levels of investments both parties proposed. The private party proposed almost

70% more of investments compared to the private party, which it wanted to spend early in 10-year period. So, also for the strategic seeking revenues plans, the private party was seen as the more aggressive.

- With respect to the fourth of the strategic dimensions (internal organisation) the intentions of the private party were seen as more opportunistic, especially with respect to its legal organisation and taking over of personnel.
- Also for the last of the strategic dimensions (external organisation) the intentions of the private party were identified as incorporating a higher level of risk. The private party relied in many of its activities on support from the Island's government, while the public party was able to operate much more autonomously.

The extent to which different institutions are conducive for the more aggressive plans of the private operators is hard to substantiate. However, several arguments can be made which make it likely that a private party is more risk-taking compared to public parties. A main argument is that in this type of 10-year project a private party will try to make as many connections and reduce the leakage as much as possible in the beginning of the contracting period in order to pick the fruits later on in the contract. Also the party will see it as a waste of money to invest large amounts of money near the end of the project as these investments will directly disappear once the project ends. Another argument is that the private party is new, and very much focused to tailor current practices with new ideas to achieve efficiency savings. Such is reflected for example in how they deal with the existing personnel.

In sum, the results from the case study points in the direction that different institutions may be conducive to formulating different strategies, acknowledging that it is hard to generalise the results from this one case.

Section III Analysis

Chapter 9 Institutions and strategic actions of WSS providers

This Chapter entails the third level of analysis on strategies, e.g. the extent to which institutions affect the strategic actions of WSS providers. In this respect a survey was conducted assessing the strategic typologies of numerous WSS providers in several countries.[51]

9.1   Introduction

The previous two Chapters analysed whether operators in a different institutional context *can* (strategic context) and *want* to (strategic plans) have different strategies. This Chapter attempts to establish whether these operators indeed also *execute* different strategies (strategic actions). Strategic actions are the realized strategies, which are assumed to have a direct relation with the performance of the operators. It is hypothesized that the institutions of WSS providers indeed make a difference for the strategies they realize.

A survey was conducted to analyse the implications of different institutional contexts on strategic actions. A sample was selected of private operators (in England & Wales), of publicly owned operators (in the Netherlands, Scotland, Northern Ireland and Italy), and of jointly owned, or mixed, operators (in Italy). The strategic actions were assessed using a typological approach as previously elaborated in the Research Design Section (see Chapter 5). A comparison is made on whether the strategic typologies of operators with different ownership characteristics provide a statistical difference.

The Chapter is structured in the following manner. First, background is given on the survey population, followed by a description of the survey methodology and outcomes. Finally the Chapter draws conclusions on the relation between institutional context and strategies.

9.2   Survey population

A survey population was selected of Italian, British and Dutch water providers. The sample population was carefully selected using a stratified sampling design by defining groups, or strata, based on ownership types. To present the spectrum of ownership possibilities, an obvious choice is to include the WSS operators from

---

[51] Parts from this Chapter have been separately published as:
- Schouten, M. and G. Casale, 2006. *Strategie dei gestori del servizio idrico in Italia.* In: Management delle Utilities. Anno-4; Numero 2. Rivesti Trimestrale. April-June 2006. Umbria.
- Schouten, M., 2007. *Exploring Strategies of Water Providers.* Paper presentation at the 1st IWA Utilities Conference 'Customer Connection'. Maastricht, the Netherlands: 14-16 June 2007.

England & Wales in the research sample. The twenty-three English and Welsh water operators are one of the few examples worldwide that are fully owned by private companies. Moreover, an advantage of inclusion of England and Wales is that there is a lot of data available on their characteristics and performance due to the ongoing yardstick benchmarking of the regulator OfWat.

To construct a proper sample it is important to identify WSS operators that are comparable to the English and Welsh water operators, but which are not privately owned. Comparability of WSS operators in terms of external environmental factors reduces the risk that data analysis is distorted. Based on this requirement the two public companies in Scotland and Northern Ireland were identified. These two operators are fully publicly owned and have relatively similar conditions under which they are operating as the English and Welsh operators.

To increase the group of publicly owned companies, also the drinking water operators in the Netherlands were included. This is quite a homogeneous group of fully publicly owned companies operating in the same Western European level of welfare and culture. And just like England and Wales, an advantage of the inclusion of the Netherlands is that there is a lot of data available on performance levels of the operators due to an ongoing benchmarking exercise by VEWIN (the association of Dutch drinking water operators).

However, the inclusion of England & Wales, Northern Ireland, Scotland and the Netherlands does not enable the research to cover the full spectrum of ownership options. Publicly owned companies on the one hand and privately owned companies on the other hand are included, but the middle group of mixed companies is absent. Hence, it is important to include another country within the sample to represent these mixed water companies. The operators within that country should also operate as much as possible in comparable conditions. In view of these requirements, also the Italian operators are included. The Italian water operators are operating in similar Western European conditions of welfare and moreover many of these operators have mixed ownership.

Although now the selection of operators is made, still it is needed to identify another level of detail, i.e. who are the informants on self-typing the strategy of an operator? The main requirement for selecting the informants within each WSS operator is to identify those employees that are in the best position to oversee and identify the realized strategies. Since strategy includes those subjects that are of primary concern to senior management, or to anyone seeking reasons for the success and failure among organizations (Joyce, 1992), informants were selected as either the general directors or the top-level management. The notion of senior management as the best-informed on strategic issues is supported by Stern and Stalk (1998: 90):

> Only the CEO can focus the entire company's attention on creating capabilities that serve customers. Only the CEO can identify and authorize the infrastructure investments on which strategies capabilities depend. Only the CEO can insulate individual managers from any-short-term penalties to the P&L of their operating units that such investments might bring about.

In each company two to three other respondents were approached to not rely only on the input of the managing director, as to avoid expression of rhetorical or intended but unrealised strategy (Andrews *et al.*, 2006). Applying such multiple informants' method is also suggested by Conant *et al.* (1990) to enhance the validity of the research. In fact, crosschecking of responses of the several respondents of the same company may help to understand whether some informants show different results than others. This has the twofold purpose to increase the final sample size and to check consistency between answers obtained inside each company. For instance, the general manager may state that the company has a certain strategic stance, because this is the formal content of the company strategy, while lower levels' managers may assess a different strategic stance based on actual observations. To broaden the group of informants for each company, the selection criterion to include an informant was that he/she was involved in strategic management decisions; usually those who were in the board of management i.e. CEO, technical director, financial director, etc.

The complete sample consisted of 191 respondents spread over 64 companies, serving more than 82 million people. For each company at least the managing director was included as a potential informant, next to 127 other informants. Of the 64 companies in the sample 25 operators were publicly owned, 23 companies had a mixed ownership, and 16 companies were privately owned. The following Table presents the distribution of WSS operators and respondents in the sample over country and ownership type. The number of –to be approached- informants is indicated between brackets behind the number of operators included in the sample.

Table 22 Survey characteristics: countries and ownership type

Sample operators	Publicly owned Companies (informants)	Mixed Companies (informants)	Privately owned Companies (informants)	*Total* Companies (informants)
England & Wales			16 (36)	*16 (36)*
Scotland	1 (2)			*1 (2)*
Northern Ireland	1 (3)			*1 (3)*
Netherlands	12 (36)			*12 (36)*
Italy	11 (36)	23 (78)		*34 (114)*
Total	**25 (77)**	**23 (78)**	**16 (36)**	*64 (191)*

Table 22 shows that publicly owned and mixed WSS operators are equally represented in the survey both in terms of number of operators as in number of informants. The privately owned operators have a lower representation in the sample, mainly because this type of ownership is rarely found in the WSS sector. The Italian operators are making up about half of the survey population, since they cover two ownership types: the publicly owned and the mixed operators.

The following Table 23 presents the distribution of WSS operators and respondents in the sample over ownership type and size. The number of respondents is indicated between brackets behind the number of operators included in the sample.

Table 23 Survey characteristics: size and ownership type

Sample operators	Publicly owned Companies (informants)	Mixed Companies (informants)	Privately owned Companies (informants)	Total Companies (informants)
Small sized (<400,000 people served)	7 (23)	11 (40)	3 (7)	21 (70)
Medium sized (between 400,000 and 1,000,000)	5 (16)	9 (28)	3 (8)	17 (52)
Large sized (>1,000,000 people served)	13 (38)	3 (10)	10 (21)	26 (69)
Total	25 (77)	23 (78)	16 (36)	64 (191)

Table 23 shows that most of the mixed (Italian) operators are relatively smaller sized compared to the publicly and privately owned companies. Overall, the distribution of company's sizes over the survey population is relatively equal. In the following sections a short description is provided of the operators that were included in the sample.

9.2.1   The participating English and Welsh WSS operators

The year 1989 marked an important turning point in the WSS sector of England and Wales. At that time the provision of WSS services in the whole of England and Wales was privatised including abstraction, production, and distribution of drinking water and the collection and treatment of the wastewater. The privatisation of the WSS sector was accompanied with the development of a new legal framework.

In England & Wales, the Department for Environment, Food and Rural Affairs is responsible for all aspects of water policy, including water supply and resources, and the regulatory systems for the water environment and the water industry. In Wales the National Assembly of Wales shoulders this responsibility. Three institutions are regulating on behalf of these institutes the drinking water providers, each from a different angle: the Office of Water Services (OfWat) is looking at the WSS providers from an economic perspective, the Drinking Water Inspectorate addresses the drinking water quality issues, while the Environmental Agency looks after the water resources.

Currently, twenty-three private companies provide regulated drinking water services for England and Wales. These companies can be classified in ten Water and Sewerage Companies and thirteen Water Supply 'only' Companies. All companies are private commercial companies with the status of so called "emanations of the state", as to force the companies to directly implement any relevant European legislation.

For the survey, all 23 prospective companies were contacted by phone. Some of the companies were contacted several times to confirm the willingness to be approached and to indicate suitable informants. 16 out of the 23 water companies were found willing to be approached for the survey and provided contact details of –to be approached- informants. The list of respondents consisted of 36 informants, which included the 16 managing directors of the companies. Hence, on average there were 2.25 informants for each water company. Table 24 below provides general data of water companies that were willing to participate in the survey. The data about these companies were gathered from several sources including OfWat, Drinking Water Inspectorate, Environmental Agency, water companies' websites supplemented by other sources.

Table 24 Participating water companies in England & Wales

#	Company	# of informants	Ownership	Water and Wastewater	Area of supply (sq. km)	Length of main (km)	# of properties (x1,000)	Population served (x1,000)	# of employees
1	Albion Water Ltd.	3	Private	No	n.a.	n.a.	n.a.	400	n.a.
2	Anglian Water Company	3	Private	Yes	27,500	36,000	2,600	4,074	3,300
3	Bournemouth & West Hampshire Water Plc	2	Private	No	1.041	2,755	192	424	176
4	Bristol Water Ltd	1	Private	No	2,400	6,400	475	1,052	403
5	Cambridge Water Company Plc	3	Private	No	1,173	2,216	134	295	119
6	Cholderton & District Water Company Ltd.	1	Private	No	n.a.	44	1	2	n.a.
7	Northumbrian Water Company	3	Private	Yes	12,261	16,930	2,000	2,441	2,427
8	Severn Trent Water Company	3	Private	Yes	20,480	45,647	3,309	7,280	4,927
9	Southern Water Services	2	Private	Yes	n.a.	13,300	1,043	2,295	2,062
10	Southwest Water	2	Private	Yes	11,137	15,000	681	1,500	1,336
11	Sutton & East Surrey Water Plc	3	Private	No	834	3,280	265	630	280
12	Tendring Hundred Water Services Ltd	3	Private	No	325	907	70	152	67
13	United Utilities	2	Private	Yes	n.a.	40,000	3,159	6,950	3,430
14	Welsh Water	2	Private	Yes	n.a.	22,202	1,200	2,900	146
15	Wessex Water	1	Private	Yes	10,000	10,500	536	1,200	1,396
16	Yorkshire Water Services	2	Private	Yes	n.a.	31,062	2,136	4,700	2,158
	Total	**36**						**36,295**	

9.2.2   The participating Scottish WSS operator

The main policy making institution related to WSS provision in Scotland is the Scottish Executive (SE). It is supported by three regulators: the Scottish Environmental Protection Agency (SEPA) for regulation of the water resources; the Drinking Water Quality Regulator, and the Water Industry Commissioner for Scotland (WIC) for the economic regulation.

Currently, one public company provides regulated drinking water services in Scotland. For the survey, Scottish Water was approached and found willing to identify two informants to be contacted for the survey, including the managing director.

Table 25 Scottish Water

#	Company	# of informants	Ownership	Water and Wastewater	Area of supply (sq. km)	Length of main (km)	# of properties (x1,000)	Population served (x1,000)	# of employees
1	Scottish Water	2	Public	Yes	78,000	46,000	2,389	5,255	4,062

9.2.3   The participating Northern Irish WSS operator

The Department of Environment is responsible for the policy making related to the Northern Irish WSS sector. Moreover, three regulators are overseeing the execution of the policies. For water resources, the Environment and Heritage Service (EHS) is responsible; for drinking water the Drinking Water Inspectorate (DWI) for Northern Ireland assumes the responsibility and as an economic regulator, the Northern Ireland Authority for Energy Regulation is acting (NIAER).

Currently, one public company provides regulated drinking water services for Northern Ireland. For the survey, the Northern Ireland Water Company was approached and found willing to identify three suitable informants to be contacted for the survey, including the managing director. See Table 26 for the general characteristics of the Northern Ireland Water Company.

Table 26 Northern Ireland Water Company

#	Company	# of informants	Ownership	Water and Wastewater	Length of main (km)	# of properties (x1,000)	Population served (x1,000)	# of employees
1	Northern Ireland Water company	3	Public	Yes	25,800	730	1,685	1,900

9.2.4   The participating Dutch WSS operators

Public administration of drinking water supply is under the responsibility of the Ministry of Physical Planning and Environment for the Drinking Water Supply Act, and the Ministry of Water Management for the abstraction on the basis of the Groundwater Act (Perdok and Wessel, 1998). The provincial and municipal government control, monitor and enforce the policies formulated by the national government. It has institutionalised a system of so-called Public Limited Companies, being a mode of organisation where the utility is incorporated as a limited company under company law, but where its stocks are owned by local, provincial, or less frequently, national government representatives. The essence of the Public Limited Company is that it uses company law as a buffer, shielding the water services business from burdensome public sector rules and regulations (Blokland *et al.*, 1999). A large part of the regulation is done via bylaws of the Public Limited Company. These bylaws, also known as the articles of association or the company constitution, are drawn up and amended by the public shareholders. The bylaws are drawn up before a public notary and need to be approved by the State for compliance with private company law, and gazetted. This means that each public limited company is differently regulated depending on the intentions of its shareholders.

The collected names and emails of potential informants of all 12 Dutch drinking water companies were to be followed up in the survey implementation phase. In total 36 informants were identified, including the 12 managing directors. Table 27 presents the general characteristics of the sample of Dutch drinking water operators.

Table 27 Participating Dutch drinking water companies

#	Company	# of informants	Owner-ship	Water and Wastewater	Length of main (km)	Annual water production (mln. m³)	Population served (x1,000)	# of employees	# of connections (x 1,000)
1	Brabant Water	3	Public	No	16,864	165	2,212	668	948
2	DZH	3	Public	No	4,430	75	1,185	456	582
3	Evides	3	Public	No	12,314	175	1,953	402	907
4	Hydron Flevoland	3	Public	No	2,249	28	307	87	126
5	Hydron Midden-Nederland	3	Public	No	6,389	76	1,231	329	557
6	Oasen	3	Public	No	4,014	45	747	221	320
7	PWN	3	Public	No	10,012	87	1,623	493	717
8	Vitens	3	Public	No	38,012	249	3,280	921	1,644
9	Waterleidingbedrijf Amsterdam	3	Public	No	2,708	88	891	520	477
10	Waterleidingbedrijf Groningen	3	Public	No	4,766	59	572	188	272
11	WMD	3	Public	No	4,216	30	432	161	191
12	WML	3	Public	No	8,466	71	1,132	426	517
Total		36			114,440	1,148	15,565	4,872	7,258

9.2.5   The participating Italian WSS operators

Recent reforms in the Italian water sector have drastically changed the organisational arrangements for the WSS provision. Historically in Italy it were the municipal administrations that carried the responsibility to provide WSS services to their populations. Hence, more than 8,000 organisations provided WSS services to the Italians, mainly in the form of the so-called *"aziende municipalizzate"*. Due to the fragmented nature of the service provision, the sector was not able to benefit sufficiently from economies of scale. The risk of small sized individual service providers choosing sub optimal solutions was apparent. Moreover, the high degree of fragmentation brought difficulties for regulatory authorities to control and monitor the performance of the service providers.

Most of the "aziende municipalizzate" struggled with limited if not poor financial capabilities. The lack of financial resources caused serious under-investments in the sector. Such was especially valid for the Southern parts of Italy as investment levels in that region were almost half of the ones employed in the North of Italy (COVIRI, 2004). The highly fragmented nature of the Italian sector and the low level of investments have been the main reasons to push for reform of the WSS sector in Italy. The objectives of the sector reform, initiated in 1994 with the enactment of Law 36 (Galli law), were to:

- Overcome the existing fragmentation of the sector;
- Achieve a higher degree of efficiency and effectiveness. Especially the objective was set to reach full cost recovery by the service providers;
- Link utilities management to integrated resources management. The Galli Law promoted the vertical integration of WSS services for a region similar to a catchment area;
- Raise customer focus and customer participation in the service provision.

The core of the reform is the definition of special geographic areas consistently with administrative constraints and river basin borders. Regional governments establish these geographical areas, which are called Optimum Territorial Areas (ATO, *Ambiti Territoriali Ottimali*). After Galli Law enactment, 91 ATOs were defined all over Italy. Parallel with the geographical definition of the ATOs, the law made provision that each area should be governed by one organisation called the Optimum Territorial Area Authority (AATO, *Autorità di ATO*). The AATO could be a consortium of municipalities (hence a newly established authority) or a convention between local governments, in which one member (for example the largest municipality) assumed the leading role. According to the intention of the reform, the AATO will be the main institutional entity that ensures water management.

Galli Law assigns to the AATO the responsibility of selecting the most suitable management option for the provider of the integrated water services (SII, *Servizio Idrico Integrato*) within the ATO, and eventually the responsibility of appointing an operator based on the selected option. Inside the ATO the law prescribes the assignment of all WSS provision responsibilities to a *single* operator. Such legal measure ensured a drastic reduction in the number of operators in Italy; theoretically from more than 8,000 to only 91. In this institutional framework, as emphasized by several observers, Galli law either aimed at or implicitly required the abolition of direct municipality management (Lobina, 2005).

Each SII operator is made responsible to provide integrated WSS services to the population living in the ATO. The law foresees different forms of institutional arrangements for the SII (Zocchi and Mangano, 2002). Either they can be fully public municipal owned companies, so called "*aziende speciali*", or they can be organised as mixed companies (SpA, *Società per Azioni*) with both public and private shareholders. It is important to note that the Galli law allowed a service provider in the form of a Joint Stock Company with majority private shareholding, provided that the assignment is in any case fulfilled through public procedures. A last option is to have private companies operating on the basis of a concession contract as SII operators.

The AATO has the role of management oversight and regulation. The AATO and the SII service provider will have to sign a Service Contract (*Convenzione di Servizio*) that functions as a tool for the AATO to define the service provider objectives and to levy penalties in case objectives are not met. In this respect the AATO plays a definitive regulative role in defining the quality of service through Service Charter (*Carta dei Servizi*), the setting of drinking water standards and the establishment of tariff levels. The institutional set-up according to Galli Law is summarized in Figure 18.

Figure 18 Italian institutional set-up based on Galli law reform

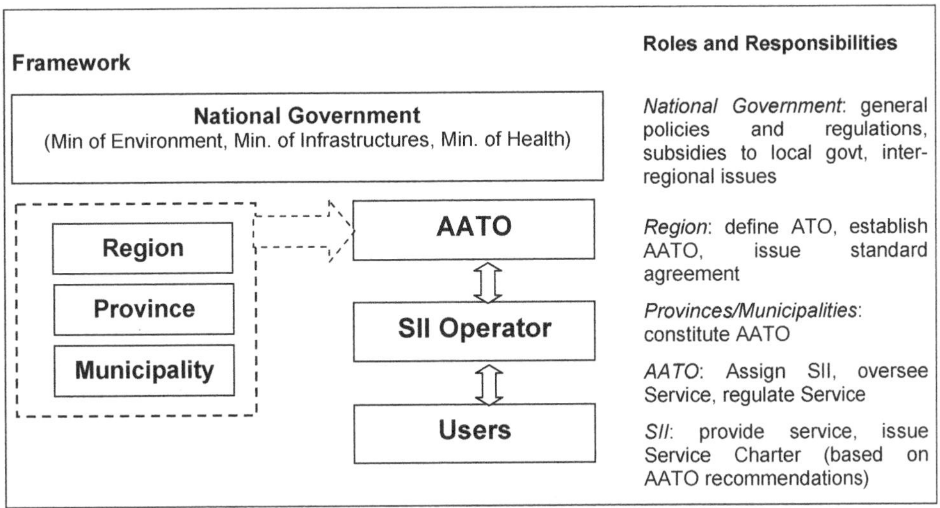

Framework	Roles and Responsibilities
National Government (Min of Environment, Min. of Infrastructures, Min. of Health)	*National Government*: general policies and regulations, subsidies to local govt, inter-regional issues
Region · **Province** · **Municipality** → **AATO** ⇕ **SII Operator** ⇕ **Users**	*Region*: define ATO, establish AATO, issue standard agreement *Provinces/Municipalities*: constitute AATO *AATO*: Assign SII, oversee Service, regulate Service *SII*: provide service, issue Service Charter (based on AATO recommendations)

Source: Zocchi and Mangano (2002). Modified by Schouten and Casale (2006).

The Galli Law defines a reform path for implementation. Once an ATO is defined by the regional authorities, the reform scheme identifies further steps to be taken. Firstly, a survey will need to be executed by the SII on the existing water supply and sanitation infrastructures (*Ricognizioni*) within the ATO region. Then an ATO Business Plans (*Piani di Ambito*) needs to be issued. Based on the Business Plan, thirdly, a SII Management Option (*Forma di Gestione Prescelta*) needs to be selected. Finally then a SII operator (*Affidamento*) can be appointed.

In 2004, the national institute overseeing the implementation of the Galli law, Coviri, assessed that 38 out of the 91 ATOs had completed the reform path by appointing a SII operator (COVIRI, 2004). The remaining ATO regions are still in a change phase. The most applied institutional arrangement in the 38 ATOs has been the one of the mixed-joint stock company. In fact 25 out of the 38 ATOs that completed the entire reform path have chosen this institutional arrangement. 12 AATOs preferred a direct assignment to an *Azienda Speciale*, de facto in the form of a fully public joint stock company, and only 1 AATO has selected a concession (COVIRI, 2004).

The focus of this research will be on the SII operators in the 38 ATO regions that completed the reform path as of 2003. In these 38 areas 40 operators are operational, as in one area (Milan province) three operators are working. Purposely the other ATO regions are excluded as strategic behaviour reflects a pattern of actions over a longer period of time. The change phase in which the other operators find themselves or the recent date of completion of the reform path, might prevent that already a clear strategic stance can be assessed. The 40 selected SII operators have different institutional arrangements and are geographically distributed (North, Centre and South of Italy). Some of the operators remained largely identical compared to the situation before Galli law, others went to substantial changes (for example the mixed joint stock companies).

The first step in contacting the 40 identified companies and potential informants was to phone each company. Telephone numbers were obtained either through the internet or through available directory through Federgasacqua (the association of public services utilities). In the final list, 34 out of the 40 companies were found willing to be approached and to provide contact details of informants. A final list of 114 contacts out of these 34 companies (meaning an average of 3.4 contacts per water operator was established.

Table 28 Participating Italian WSS providers

	Company	# of informants	Owner -ship	Multi utility	Population served (x1,000)	# of municipalities	Area of supply (sq. km)
1	ACA S.p.A.	4	Public	No	436	64	1,732
2	ACEA S.p.A.	2	Mixed	Yes	3,599	112	5,109
3	Acqualatina S.p.A.	4	Mixed	No	574	38	2,498
4	Acque del Chiampo S.p.A.	1	Mixed	No	54	10	162
5	Acquedotto del Fiora S.p.A.	6	Mixed	No	378	55	7,484
6	Acquedotto Lucano S.p.A.	3	Mixed	No	597	131	9,992
7	Acque S.p.A.	5	Mixed	No	753	60	3,125
8	AEMME Acqua S.p.A.	2	Mixed	No	754	68	600
9	AMAG S.p.A	2	Public	Yes	311	147	2,810
10	AQP S.p.A.	3	Public	No	4,019	258	19,363
11	ASM Brescia S.p.A.	4	Public	Yes	1,108	206	4,784
12	ASP S.p.A.	4	Mixed	Yes	253	154	2,015
13	BIM Gestione Servizi Pubblici S.p.A.	3	Public	Yes	203	66	3,596
14	Brianza Acque S.p.A.	1	Mixed	No	814	70	600
15	CAM S.p.A.	3	Public	No	127	35	1,764
16	CIIP S.p.A.	4	Mixed	No	288	59	1,813
17	Genova Acque S.p.A.	3	Mixed	No	878	67	1,838
18	Gorgovivo Multiservizi S.p.A	5	Public	Yes	387	45	1,816
19	GORI S.p.A.	4	Mixed	No	1,468	78	906
20	Gran Sasso Acque S.p.A.	4	Mixed	No	100	37	1,803
21	HERA S.p.A.	2	Mixed	Yes	272	20	534
22	Metropolitana Milanese S.p.A.	2	Public	Yes	1,256	1	182
23	Miacqua S.p.A.	2	Mixed	No	875	51	600
24	Nuove Acque S.p.A.	6	Mixed	No	300	37	3,262
25	Publiacqua S.p.A.	4	Mixed	No	1,191	50	3,727
26	Russo Servizi S.p.A.	3	Public	n.a.	253	40	1,701
27	SACA S.p.A.	3	Mixed	n.a.	75	37	1,502
28	Salerno Sistemi S.p.A.	4	Mixed	No	773	144	4,763
29	S.A.S.I. S.p.A.	2	Mixed	No	270	92	2,298
30	Servizio Idrico Integrato Terni	4	Mixed	No	217	32	1,953
31	SMAT S.p.A.	3	Public	No	2,153	306	6,713
32	Tennacola S.p.A.	4	Public	No	114	27	652
33	Umbra Acque S.p.A.	4	Mixed	No	457	38	4,302
34	Valle Umbra Servizi S.p.A.	4	Mixed	Yes	151	22	2,201
Total		*114*			**25,458**	**2,657**	**108.200**

Since the Italian WSS providers are the only ones in the survey that include the mixed type (or jointly owned type); it may be valuable to elaborate further on this institutional arrangement. In the Table above for a number of operators the public shares in the stock of the mixed operators is provided (Marra, 2006). Although the number is small, it indicates the general trend in which total public shares are more than 50%. By retaining the majority of the stock, the public entity is able to gather information about actual management and have a greater say on the board of the operator. Majority ownership gives proprietary information to the government to stimulate the provider to lower operating costs and monitor it more effectively with respect to the fulfilment of contractual obligations.

9.3   Survey methodology and response

The survey methodology is based on Dillman (2000), who developed the -so called-Tailored Design Method (TDM). After the questionnaire was constructed, it was several times revised based on a series of pre-testing efforts. The questionnaire was pre-tested in fora of knowledgeable colleagues, outsiders to the research, and members from the target population. Their input greatly improved the structure and content of the questionnaire[52]. The Dutch version of the questionnaire is added in Annex 6. Apart from the Dutch version, also an English and an Italian version were developed for the respective survey populations.

The survey implementation ended after five contacts with each of the informants using the step-by-step survey methodology suggested by Dillman (2000). Table 29 below shows the summary of responses to the survey divided over the types of ownership and the countries of the operators. The number of respondents is put between brackets after the number of operators that responded.

Table 29 Number of responses divided over country and ownership type

# of responses Ownership/ country	Publicly owned Companies (respondents)	Mixed Companies (respondents)	Privately owned Companies (respondents)	Anonymous (respondent)	Total Companies (respondent)
England & Wales			10 (17)		10 (13)
Scotland	1 (1)			(1)	1 (1)
Northern Ireland	1 (3)				1 (3)
Netherlands	10 (23)			(2)	10 (25)
Italy	11 (19)	16 (28)		(2)	27 (49)
Total	**23 (48)**	**16 (28)**	**10 (17)**	**(5)**	**49 (96)**

The above Table shows that 96 respondents have responded to the survey distributed over 49 water operators. Five anonymous responses were received. Since each of the questionnaires was set up in the Italian, Dutch or English language, it was possible to determine the country of origin of the anonymous replies. Most of the responses were received from the publicly owned operators and from the Italian water operators. The below Table 30 shows that such is partly due to the relatively higher portion of these

---

[52] Pre-testing interviews were conducted with A. Freijters, Manager for Operation and Water Technology of PWN, Mr. Van den Boogaard and J. Hoffer, both managers from Vitens-Evides International, H. de Jonge, Head of Strategy and Policy of DZH.

types of operators in the sample, but also due to their high response rates. The overall response rate at company level was 77% and at respondent level was 50% which were deemed satisfactory.

Table 30 Response rate divided over country and ownership type

Response rate Ownership/country	Publicly owned Companies (respondent)	Mixed Companies (respondent)	Privately owned Companies (respondent)	Total Companies (respondent)
England & Wales			63% (47%)	63% (50%)
Scotland	100% (50%)			100% (50%)
Northern Ireland	100% (100%)			100% (100%)
Netherlands	83% (64%)			83% (69%)
Italy	100% (53%)	70% (36%)		79% (43%)
Total	**92% (60%)**	**70% (36%)**	**63% (47%)**	**77% (50%)**

It can be noted, by assessing the responses relative to size of the operators, that the distribution of operators is they are relatively equally distributed (see right column of Table 31), although medium sized companies are relatively under-represented. However, it is to be noted that the mixed companies participating in the survey are clearly smaller sized compared to the other two groups of operators.

Table 31 Number of responses divided over size and ownership type

# of responses Ownership/ size	Publicly owned Companies (respondent)	Mixed Companies (respondent)	Privately owned Companies (respondent)	Anonymous (respondent)	Total Companies (respondent)
Small sized	7 (9)	9 (18)	1 (1)		17 (28)
Medium sized	3 (8)	6 (9)	3 (7)		12 (24)
Large sized	13 (29)	1 (1)	6 (9)		20 (39)
Anonymous	(2)			(3)	(5)
Total	**23 (48)**	**16 (28)**	**10 (17)**	**(3)**	**49 (96)**

The response rates of the survey, allocated over the different sizes and ownership types of the operators is presented in the below Table 32. It shows that the response rate of large operators is fairly smaller compared to the responses of the smaller sized companies, although at respondent-level the response rates are almost similar.

Table 32 Response rate divided over size and ownership type

Response rate Ownership/ size	Publicly owned Companies (respondents)	Mixed Companies (respondents)	Privately owned Companies (respondents)	Total Companies (respondents)
Small sized	100% (44%)	82% (45%)	33% (14%)	80% (42%)
Medium sized	75% (46%)	67% (32%)	100% (90%)	77% (47%)
Large sized	100% (81%)	33% (10%)	50% (37%)	73% (55%)
Total	**92% (62%)**	**63% (36%)**	**92% (47%)**	**77% (50%)**

9.4   Analysis of self-description and aggregated multi-item scores

The last question in the questionnaire was based on the paragraph description methodology (see Chapter 5). The question was formulated as follows: *"In which of the below descriptions do you find the most similarities with your own water company? Please tick the appropriate box (only one)"*. The question was followed by a description of 4 different types of companies (similar to the description provided in Table 4). Each of the four descriptions reflected one of the four strategic typologies (Type A is Prospector, Type B is Defender, Type C is Analyser, and Type D is Reactor).

The Graph below investigates the responses from the 96 respondents, distributed over ownership type.

Graph 3 Responses on self-description

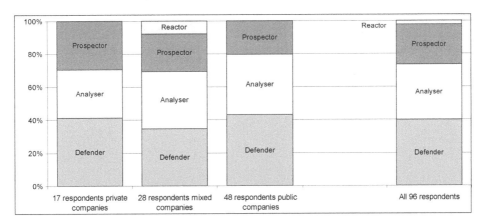

The data shows clearly that only a small amount of all respondents perceive their company to be a Reactor (only 2 responses). Both of these respondents were from mixed companies. Most of respondents (38%) indicated their company to be a Defender, followed by Analyser (31%) and Prospector (23%), which is shown in the column on the far right. The three columns on the left, representing the responses categorized per ownership type, are fairly similar. The major difference is that respondents from private companies respond comparatively more to be a Prospector.

To analyse the consistency of responses, it is assessed whether the self-description is the same from respondents of the same company. Analysis shows that for a small majority (52%) the responses for the self-description are similar of informants from the same company. Such may suggest either that the there is little consensus within companies on the strategies pursued, or that the self-description is a not very precise measuring instrument.

Another analysis that helps to check the consistency of the responses is to compare the outcomes of the paragraph description with the results of the multi-item description. In that respect aggregated scores are made at respondent level for all 22 variables both for the Defender-Prospector scale, and the Non-Reactor – Reactor scale. In Annex 7 a more in depth comparison is presented between the self-description and the multi-item measurement.

The scores on the aggregated Prospector-Defender scale range from a minimum score of 33 points to a maximum score of 68 points per respondent. To classify the companies into one of the typologies, the highest tertile is classified as Prospectors, the lowest tertile as Defenders, while the middle group is classified as Analysers. The scores on the aggregated Non Reactor-Reactor scale range from a minimum score of 21 points, to a maximum score per respondent of 66 points. All score above the median score of 44 points are classified as being Reactors. The result of this exercise is presented in Graph 4:

Graph 4 Aggregated scores on multi-item measurement

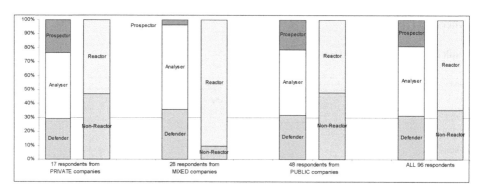

Graph 4 differs from the previously presented Graph 3 since in the multi-item measurement two Likert-scales are incorporated (the Prospector-scale and the Reactor-scale), while the self-description exists of a non-ordinal scale of 4 strategic archetypes.

The six columns on the left of the Graph present the different scores distributed over ownership type. Interestingly, while the scores of the private and the public companies almost show a similar distribution, the scores of the mixed companies are quite different. Prospectors are largely under-represented in this group, and the group has strong Reactor-like characteristics. Such is partially validated by comparing it to the outcome of the self-description in Graph 3. Also in that case Prospectors were relatively under-represented for the mixed companies, and the mixed companies were indeed the only category which included companies from the Reactor category.

The data included in Graph 3 and Graph 4 is based on the responses from all 96 responses. To understand whether respondents from one company are consistent in their answers, it is analysed whether the responses within one company are similar. 70 on the responses were from respondents working for the same companies. Several analytical steps are conducted in this regard. The calculation shows that for a much larger percentage (85%) the outcomes of the aggregated multi-item score are similar from respondents from the same company. In other words, in 85% of the cases the aggregated score of multi-item test was the same for two respondents from one company. This result suggest that on the one hand the multi-item test is a more precise tool to define exactly the typology, compared to the paragraph description; while on the other hand it suggests that strategies may indeed be well understood within one company and broadly supported.

The above analysis focussed on the outcomes of the aggregated scores, however by making an inter-rater analysis it can also be assessed whether the responses are uniform for each of the 45 questions in the questionnaire. This analysis is important as it may justify using a mean of more than 1 response per company. The inter-rater analysis entails a comparison between the average correlation coefficient of the whole survey population, and the average correlation coefficient of the group which has more than 1 response per company. The unit of analysis is the operator and the different respondents work as "samples" coming from the same unit of analysis. The average correlation coefficient of the total of responses by the 96 respondents given to the 45 questions (44 questions on the multi-item scale and the one question for the paragraph description) is calculated at 0.42 (excluding the variables with missing data). This correlation coefficient of the total of responses indicates that the responses are positively correlated (the scale of correlation coefficient is from -1 to +1). Within the group of 28 operators the average number of responses was 2.5 responses per operator. A calculation was made to determine the correlation coefficient of the responses within each of these 28 operators. The average of the 28 correlation coefficients was calculated at a value of 0.55. This value indicates that the responses were to a higher degree correlated compared to the responses of all of the respondents. This higher correlation indicates that the responses of respondents that work within one operator are more uniform compared to the ones that are working with separate companies. Based on this result, aggregation of operators' respondents was performed by using average of each variable.

Another valuable analysis step to be made is to check the internal consistency of the two scales. Hence to what extent is it justified to aggregate the scores of individual questions to an overall score. Within the questionnaire 22 questions provide a scoring for the Defender-Analyser-Prospector scale; while another 22 questions provide scores for the Non Reactor-Reactor scale. To establish whether aggregation of individual scores is justified Cronbach's coefficient alpha is calculated. A survey's internal consistency, or homogeneity, is the extent to which all the variables (or questions) assess the same, skill, characteristic, or quality (Fink, 1995). Cronbach's coefficient alpha, the average of all the correlations between each variable and the total score, is often calculated to determine the extent of homogeneity between the variables measured.

On the Prospector-Defender-Defender scale a Cronbach Alpha value was calculated at 0.48, while for the variables related to the Reactor Stance a Cronbach Alpha value of 0.71 was calculated. These values indicate a high consistence of the variables to assess the Reactor Stance. This indicates that for both scales there is positive consistency in the responses, although the internal consistency on the Reactor-scale is much higher. Meaning, the variables used to assess the extent to which operators score on the Non Reactor-Reactor continuum are more correlated, compared to the variables used to assess the extent to which operators score on the Prospector-Defender continuum. Based on this analysis it was found justified to aggregate the variables for both scales, although specific attention is given to this issue by calculating also for each individual strategic action (markets, products/services, seeking revenues, internal and external organisation) the Cronbach alpha values.

Some observations can be made comparing Graph 3 and Graph 4. Looking at the columns on the right one can see that the distribution over Prospector, Analyser and Defender are relatively similar, although the multi-item analysis shows relative more

Analysers compared to Defenders. This relative similarity is partially confirmed by the calculation that 52% of typologies from the self-description and the multi-item test are equal to one another. Hence, using the output from both the output of the self-description and of the aggregated multi-item test it is concluded that overall in the WSS sector the dominant typologies are Analyser and Defender. Prospectors are more dominant in the private group of companies, while the mixed companies have more Reactor characteristics. Moreover, the outcome strengthens the notion that strategies should be assessed at the lower strategic action levels (i.e. markets, services, seeking revenues, internal organisation and external organisation).

9.5   Analysis for each of the 5 strategic components

The following sections will address each of the strategic actions, and will identify for each strategic action which of the variables showed a statistically significant difference between ownership types. Analysis of variance (ANOVA) is used to test hypotheses about differences between two or more means. Box 4 below provides background on the ANOVA methodology.

Box 4 ANOVA analysis

ANOVA assumes that each group is an independent random sample from a normal population. Analysis of variance is robust to departures from normality, although the data should be symmetric. When there are more than two means, it is possible to compare each mean with each other mean using t-tests. ANOVA validates which variables really give significant difference. For the statistical analysis both the uni-variate F-value and the p (probability) value are calculated.

The key statistic in ANOVA is the F-test of difference of group means, testing if the means of the groups formed by values of the independent variable (or combinations of values for multiple independent variables) are different enough not to have occurred by chance. The uni-variate F-test is to analyse differences in means for statistical significance. The F test analyses whether the ratio of the two variance estimates is significantly greater than 1. Hence, the larger the F-value, the more the groups differ from another. If the group means do not differ significantly then it is inferred that the independent variable(s) did not have an effect on the dependent variable.

The p-value represents the probability of error that is involved in accepting the observed result as valid, that is, as "representative of the population". Traditionally, experimenters have used either the 0.05 level (sometimes called the 5% level) or the .01 level (1% level), although the choice of levels is largely subjective. The lower the significance level, the more the data must diverge from the null hypothesis to be significant. In this research minimum significance level was set at 0.05 (5%) level. For variables that give significance value larger than 0.05, its hypotheses will be rejected. It means that the difference between two groups is not significant enough. A p-value of 0.05 (i.e. 1/20) indicates that there is a 5% probability that the relation between the variables found in our sample is a "fluke." The statistical significance of the comparison of the average values of each group of operators is the probability that the observed difference occurred by pure chance ("luck of the draw"), and that in the population from which the sample was drawn, no such relationship or differences exist.

Annex 8 gives an overview of the outcome of the statistical analysis of the data. For the analysis a further elaboration is made in the following sections, using the scatter diagram from the analytical framework developed in Chapter 4.

9.5.1   Strategic market actions

In the Research Design section (specifically Chapter 4) the four variables used to assess the market strategic actions were identified. These were: connections, inset appointments, bulk water, and mergers and acquisitions. For each of these variables two questions were included in the questionnaire: one to measure the extent to which an organisation scores on the Defender-Prospector scale, and one to assess the extent to which a provider has Reactor characteristics. The aggregated result of the 4 variables is presented in the below scatter box diagram:

Scatter Box 2 Aggregated scores strategic market actions

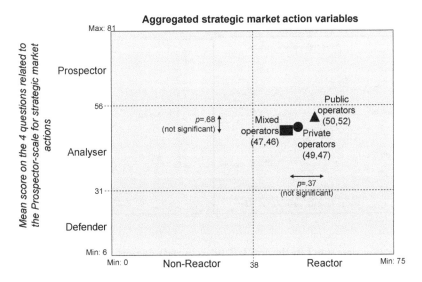

Mean score on 4 questions related to the Reactor-scale for strategic market questions

The scatter box diagram shows that all three types of companies are positioned in the middle of the scatter box, suggesting an Analyser typology, with moderate Reactor characteristics. The Graph also shows that the differences between the three types of companies are not significant.

Cronbach's Alpha is calculated for both scales to analyse the internal consistency of the responses for the strategic market actions. For the Defender-Analyser-Prospector scale the Cronbach Alpha value was calculated at minus 0.03, which suggest inconsistency in the responses per question. For the Reactor scale, the Cronbach alpha value suggested a much higher consistency (value at positive 0.45), but still this is statistically not high enough to validate the scale.

Going one level deeper, it is analysed to what extent there are differences in the responses for each of the four variables that make up the aggregated score. Making this analysis, it is found that for three out of the four variables indeed there is a statistically significant difference between the three groups of companies. Only for the 'bulk water' variable no statistically significant difference was found (see Scatter Box 3 below).

Scatter Box 3 Relative positions for the four individual strategic market variables

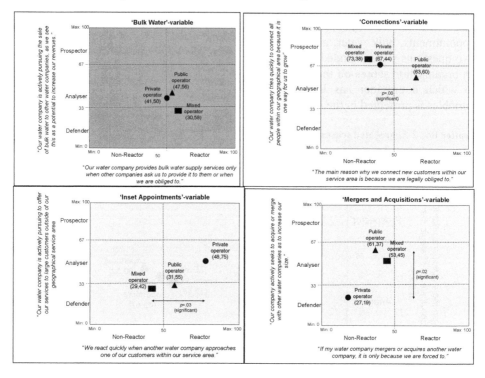

The scatter box presented in the top left box (in red) depicts the relative positions of the three companies for the 'bulk water variable'. It shows that the three types of companies score relatively similar on the strategic market actions. In this respect, the typology is Analyser, with moderate Reactor characteristics.

For the other three variables statistically significant differences were found, i.e. for the Reactor-scales of 'connections' and 'inset appointments', and the Prospector scale of 'mergers and acquisitions'. The other boxes present these significant differences. Analysis of the variable of 'connections' (represented in the top right scatter box) shows that all companies have a relatively strong Prospector behaviour. Hence, all of the companies say they try to quickly connect all people that live within their geographical areas. However, on the Reactor scale for this variable a statistically significant difference is found. The responses from public companies differ statistically from the responses of mixed and private operators. Public companies agreed significantly more to the statement *"The main reason why we connect new customers within our service area is because we are legally obliged to"*. This more Reactor type of behaviour of publicly owned WSS operators may be because private and mixed companies do not need much motivation to connect new customers from the government, and will in their response not indicate that the legal obligation is the prime determinant. In view of their profit objective, the private and mixed companies may be motivated to connect new customers as they serve as a source for more revenues.

The results of the analysis for the variable of 'inset appointments' are presented in the lower left scatter box. The Graph shows that all companies have an Analyser type of behaviour, but that on the Reactor scale the three types of companies give different responses. Private companies are the most Reactor like, and mixed companies the least Reactor like. This means that private companies agreed significantly more to the statement "*We react quickly when another water undertaker approaches on of our customers within our service area*"; compared to the public and mixed companies. This more Reactor type of behaviour of privately owned WSS operators can be explained by several factors.

- One factor is the higher dependency of private operators on the revenue potential of the large customers within their service area, making it more important for them to keep them as customers.

- Another factor is that in case neighbouring water companies are privately owned, there is a higher likeliness they may actively approach customers outside of their service areas. This factor is partly supported by the findings on the Prospector scale of inset appointments, in which respondent were asked to identify whether their companies actively approach customers outside of their service area. Private companies score higher on this question, although the difference with the other companies was not statistically significant.

- A third factor explaining the higher score of private companies on the Reactor scale of 'inset appointments' is connected to the high importance given by the English regulator on inset appointments, giving a higher contestability of the market for large customers. This factor is supported by making an analysis based on country of origin[53]. If an analysis is made along countries of origin, the responses from England & Wales are statistically significantly higher from the other responses (p=0.009). Scotland and Northern Ireland is this analysis score much lower on the Reactor-scale compared to responses from operators in other countries. An explanation for this low Reactor-score of the Scottish and Northern Irish operators may be explained by the fact that both operators in these countries have the entire country as their service area.

The scatter box presented at the lower right of the Scatter Box 3 presents the results of the analysis for the variable of 'mergers and acquisitions'. A statistically significant difference was found on the Prospector scale. The more public the ownership of the company, the more the respondents agreed to the statement: "*My company actively seeks to acquire or merge with other water companies as to increase our size*", Looking at the scatter box it is clear that the private operator gives very different responses compared to the other two groups of company types. The private operator is more Defender like, but also less Reactor compared to other two. The different responses from the private companies can be explained by looking at the country specifics. The Dutch water companies have gone through a series of mergers in the last decades bringing down the number of water providers from several hundred to only fourteen in 2008. Expectation is that this development will continue until only 3 or 4 companies remain. For the Italian providers to merge with other companies is

---

[53] Checking whether also size of the operators may be factor accounting for the difference for operators of different ownership shows that a grouping of operators along size provides no significant difference.

inherent to implementation of the Galli Law, in which one provider is made responsible for a whole catchment area. Hence, considering these country specifics both the Dutch and the Italian provider will agree more to this statement. On the other hand, the English and Welsh providers will score much lower on this variable. The regulator OfWat discourages water companies to further merge with other providers within the English territory as they want to maintain a reasonable amount of companies to be included in the yardstick benchmark. These findings go hand in hand with the finding if an analysis per country is executed[54]. The Dutch water providers were in this case the most active in seeking opportunities to merge with, or acquire, other water providers, while the UK operators were the most inactive in this respect (*p*=0.000 for Prospector).

9.5.2   Strategic products/services actions

Two variables were identified in Chapter 4 as a means for assessing the strategic archetype of products and service, e.g. 'quality' and 'product portfolio'. For the aggregated scores on this strategic action no statistically significant difference was found (see below Scatter Box 4):

Scatter Box 4 Aggregated scores for the strategic products and service actions

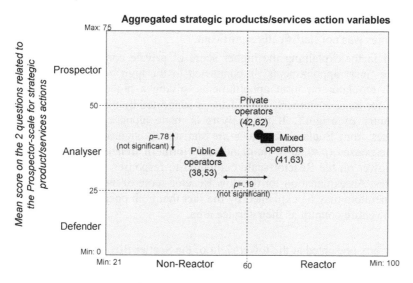

Mean score on 2 questions related to the Reactor-scale for strategic product/services questions

The above Scatter Box 4 shows that the aggregated scores for all three types of companies are very similar to one another. All companies can be classified as Analysers. Public operators are less Reactor-like compared to the other two types, although the difference is not statistically significant.

The internal consistency of the scores on the strategic product/services action is calculated through Cronbach Alpha value. The Cronbach alpha value for the Prospector scale amounted to 0.54, while it was calculated for the Reactor scale at

---

[54] Checking whether also size of the operators may be factor accounting for the difference for operators of different ownership shows that a grouping of operators along size provides no significant difference.

minus 0.21. Hence, for the Prospector scale the responses showed much more consistency although using the criteria of Cronbach Alpha still this is considered 'poor' internal consistency.

Again investigating the extent to which differences occur at the level of individual variables, it is found that for one variable a statistically significant difference was found e.g. for the 'quality'-variable. For the variable of product portfolio no significant was found (see Scatter Box 5).

Scatter Box 5 Relative positions for the strategic products/services variables

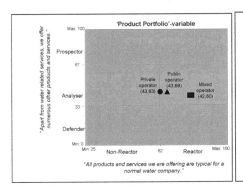
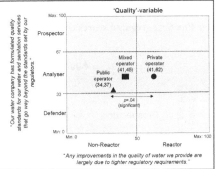

The Scatter Box shows that the responses for the 'Product Portfolio'-variable on the Prospector scale were for this variable almost identical for all three groups of companies ($p=.99$). This outcome suggests Analyser behaviour with respect to whether the companies also offer other products and services apart from water. On the Reactor scale there is some variation between the answers of the different companies, but the differences are not statistically significant. However, the Graph shows that all three companies have a relative strong Reactor-type of behaviour for the variable.

For the other variable ('quality') composing the strategic product and service actions a statistically significant difference was found (see scatter box on the right side). From the presented scatter diagram, it can be derived that on the Prospector scale all operators score very similar, at an Analyser level. For Reactor scale the responses are more distributed. The data would suggest that the more private the company, the more the respondent agreed to the statement: "*Any improvements in the quality of the water we provide are largely due to tighter regulatory requirements*".[55] The statistical difference cannot alone be attributed to the ownership structure, but should be combined with the difference in regulatory environments. Related to the relation between regulatory environments and the combination of the operator's strategies towards changing services a statistical significant difference was found on the Reactor scale, i.e. $p<0.05$. The Dutch are the least Reactor like. An explanation might be that the Dutch water operators are serving their customers with water quality that exceeds the legal mandatory quality levels. Hence, they are trying to serve their existing customers with the best possible water, even though such is not legally mandatory. Since the Dutch water quality is better than the legally required water quality they are less Reactor-like in their behaviour towards external pressures from regulators. The operators in the other countries are more inclined to adhere to the regulations related

---

[55] In the questionnaire the "regulatory requirements" were made specific for each country context.

to water quality. England & Wales and Italy score fairly similar in respect to the Reactor-like behaviour. Company size is not an explanatory variable for the difference strategies related to ownership.

9.5.3   Strategic seeking revenues actions

Four variables were selected in Chapter 4 of the Research Design section to measure the strategies of the sample group of operators in their pursuit to seek revenues, e.g. how they allocate their profits, the extent they seek to realize efficiency improvements, the extent they conduct asset management, and whether they are willing to provide water-for-free as a gratitude. The aggregated scores for the 5 variables are depicted in the below Scatter Box 6:

Scatter Box 6 Aggregated scores on the strategic 'seeking revenues' action

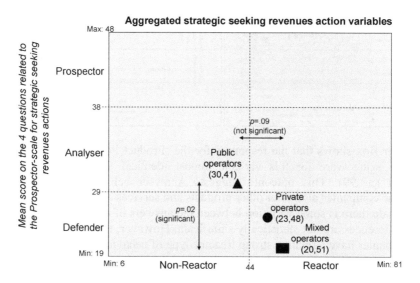

Mean score on 4 questions related to the Reactor-
scale for strategic seeking revenues questions

The Scatter Box shows that the private and the mixed companies can be classified as Defenders, while the public companies are Analysers. The responses from the public companies are in this respect statistically different ($p=.02$), On the Reactor scale no significant difference was found, although also in this case it is observed that public companies are distanced from the other two, suggesting that the public companies are less Reactor-like.

The Cronbach Alpha value is calculated to assess the internal consistency of the responses with respect to the strategic seeking revenue action. For the Prospector scale the Cronbach Alpha was calculated at 0.54, and for the Reactor scale at 0.26. Hence, using the Cronbach alpha criteria for both scales the internal consistency is low.

Investigating at the level of individual variables for three out of the four variables statistically significant differences were found. Only for the variable of asset management the respondents from public, mixed and private companies responded in a similar way (see Scatter Box 7):

Scatter Box 7 Relative positions for the strategic 'seeking revenues' variables

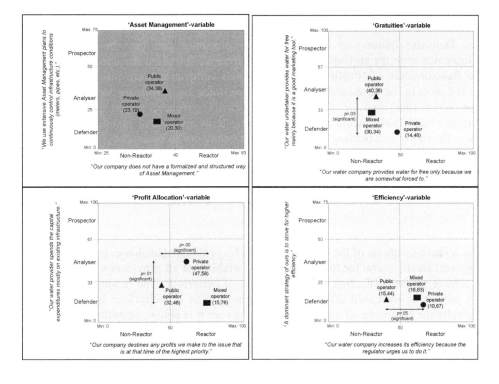

The above Scatter Box presented in the top left, shows that all companies score relative low on both scales for the 'Asset management'-variable. Such would suggest that the companies have a Defender attitude, combined with a Non-Reactor attitude.

For three variables statistically significant differences were found (see the blue scatter boxes). The high score of public companies can for a large part account for the statistical difference on the consolidated Prospector-scale on the variable 'Gratuities' (see top right scatter box diagram). Respondents from public companies agreed more to the statement: "*Our water undertaker provides water for free mainly because it is a good marketing tool*". Publicly owned companies value gratuities more compared to companies with private sector involvement. Such seems relatively logical, as shareholders from private companies are governmental entities (like national governments and municipalities), which have an interest to make water available to their citizens. Private shareholders do not have such incentive. It might be for their profit-oriented nature that fully private owned operators are least prone to provide water for free, unless they are obliged to because of regulatory interferences.

Also for the variable of 'Profit allocation' a statistical significant difference was found (see scatter box diagram on the lower left). On the Prospector scale respondents from mixed companies agreed the most, and respondents from private companies the least, with the statement "*Our water company spends the capital expenditures mostly on existing infrastructure*". Such can be explained by the necessity for Italian mixed (and public) providers to invest relatively heavily in maintaining the existing infrastructure in view of the oftentimes poor state. The current state of infrastructure in the Netherlands and the UK is much better, leaving more money for these providers to invest in new ventures.

The Defender attitude of mixed (Italian) companies can best be interpreted in conjunction with the statistical significant difference found on the question related to the Reactor-scale of 'Profit allocation'. Mixed companies were also the ones that agreed most to the statement "*Our company destines any profits we make to the issue that is at that time of the highest priority.*" Hence, mixed companies are comparatively fairly inconsistent in their long term investment planning. They have taken flexibility in adjusting their investment behaviour to emergency needs, possibly due the sudden demands from infrastructure collapses. In the Netherlands and the UK major unforeseen collapses in their infrastructure are less frequent; hence these companies are more able to pursue a steady long term strategy. Italian providers, on the other hand, are forced to make an annual reassessment of the highest priority due to the unreliability of their current assets.

The relative positions of the three types of WSS providers (see lower right scatter box diagram) clearly show for the 'Efficiency'-variable that all providers value efficiency to a large extent, indicating a Defender type of behaviour. The rationale to interpret the pursuit of increased efficiency as Defender type of behaviour is because it is typically the behaviour of a company to make sure it is able to consolidate its current position by making better use of its current resources. A statistical significant difference was found on the Reactor-scale. Private companies agreed more to the statement that: "*Our water company increases its efficiency because the regulators urge us to do it.*" Private operators perceive a higher pressure from regulators to increase their efficiency, while publicly owned operators also pursue efficiency improvement but they are more internally motivated. Such may be accounted to the more active role of regulators of private companies in giving directions with respect to efficiency levels (for example the English price-cap system).

9.5.4   Strategic internal organisation actions

Seven variables were selected in Chapter 5 to measure the strategies of the sample group of operators related to their internal organisation, e.g. how they embrace innovations, award multidisciplinary knowledge and skills, encourage training, make their strategies, view (de) centralisation, empower their employees, and strategize their marketing actions. The aggregated data analysis for the 7 identified variables results in Scatter Box 8.

Scatter Box 8 Aggregated strategic internal organisation action-variables

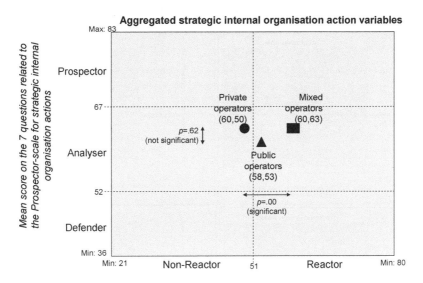

Aggregated strategic internal organisation action variables

Mean score on 7 questions related to the Reactor-scale for strategic internal organisation questions

For the aggregated variables measuring the strategic stance towards Internal Organisation a statistically significant difference was found on the Reactor scale. Private companies (English and Welsh) are the least Reactor-like, while mixed (Italian) companies have the most Reactor-characteristics. Such may be explained by lower levels of managerial discretion in the case when shares are kept by both public and private parties. It may be that due to the constant interaction between public and private shareholders in these firms the (visibility of) interferences of these shareholders is higher. Interestingly, analysing the data based on company size a statistically significant different is found for the consolidated variables on the Reactor scale. The smaller the company size, the more the respondents answered Reactor-like to the statements on 'Internal Organisation'. Apparently the smaller a company is, the more it is prone to external influences in their strategies towards internal organisation.

Also for the strategic internal organisation action the internal consistency was assessed by calculating the Cronbach Alpha value. For the Prospector scale the Cronbach Alpha was calculated at 0.56 and for the Reactor scale at 0.54. Hence, internal consistency at both scales is judged as 'poor' using the criteria of Cronbach alpha. Looking more in detail, for four variables no statistically significant differences were found (see Scatter Box 9).

Scatter Box 9 Non significant differences for the 'innovation', 'training', 'decentralisation' and 'marketing' variables of the strategic internal organisation action

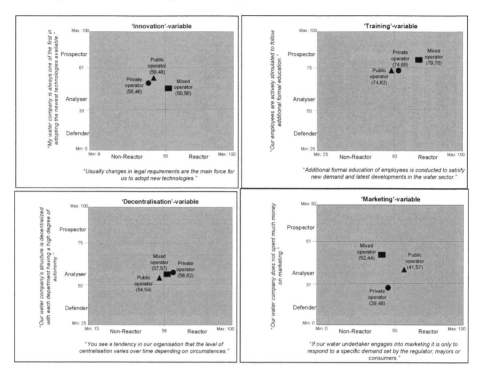

Investigating the four variables with non significant differences, it shows that all of them score approximately in the middle of the Reactor scale. Most of them also score in the middle of the Prospector-scale, suggesting Analyser behaviour. Only the variable of 'Training' depicted in the top right scatter box gives Prospector behaviour for all three groups of companies. Hence, all companies agreed very much to the statement that their employees are actively stimulated to follow additional formal education.

For the remaining three variables composing the strategic internal organisation actions, statistically significant differences were found, i.e. for the 'knowledge and skills', 'strategy-making' and 'empowerment' variables (see the 3 scatter box diagrams below).

Scatter Box 10 Statistically significant differences for the strategic internal organisation variables

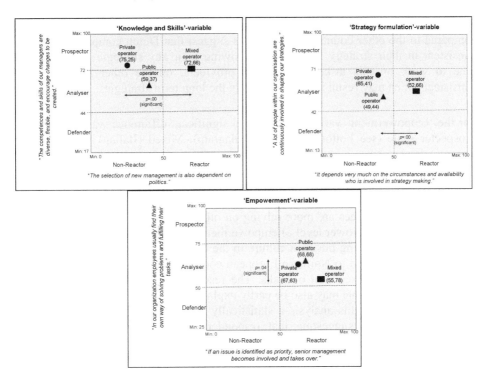

For the variable of 'Knowledge and Skills', represented in top left scatter box a significant difference was found on the Reactor scale. Respondents from mixed companies agreed comparatively more to the statement *"The selection of new management is also dependent on politics"*. Hence, the interference of external parties on the selection of the management of the operator is highest for mixed (Italian) companies. The Dutch respondents scored the lowest on this statement. The Dutch are the least depended on external parties in the selection of their managers. This result seems not really logical. One would perceive that a fully private company, like the water providers in the UK, would have less interference from politicians, than the public companies in the Netherlands. One can notice for example, that in many instances the managing directors of Dutch water providers are former politicians. A possible explanation for the lower score of the UK company on this variable may be that the respondents interpreted the term 'politics' in the question broader, by also including internal company politics. In such case their lower score makes sense, as in many cases the UK water undertakers are part of large multinational firms, in which the managing director position of the daughter company (the water undertaker) is one amongst many.

Also for the variable of 'Strategic formulation' a statistical significant difference was found on the Reactor scale (see scatter box on the top right). Again mixed (Italian) companies showed the most Reactor-type of behaviour, by agreeing most to the statement *"It depends very much on the circumstances and availability who is involved in strategy making"*. The data suggest that in mixed (Italian) companies the extent to who needs to be involved in the strategy making process is more inconsistent and depending on availability of employees compared to the situation in the publicly owned and privately owned companies. Just like for the 'knowledge and skills'-variable, the Dutch are the least Reactor-like, meaning that they are more consistent in the selection of employees that need to participate in the strategy making process compared to the other countries. This results suggest that Dutch companies are more consistent in their strategy formulation; allowing their key strategic employees the time to work on this task. Italian water providers, on the other hand, are less consistent which may result in annual shifts in the long term strategies.

For the 'Empowerment' variable a statistical significant difference was found on the Prospector scale (see bottom scatter box diagram). Mixed (Italian) companies are more Defender like compared to public and privately owned companies in empowering their employees. Such is deducted from their lower scores on the statement: "*In our organisation employees usually find their own way of solving problems and fulfilling tasks*". The data suggests that in companies with mixed ownership the employees are more relying on others in the execution on their tasks. An explanation of the lower level of empowerment in Italian water provider may also be related to the reigning business culture in the country. It may be that in Italy it is less common compared to Northern European countries for employees to enjoy a lot of freedom in their execution of tasks. The result on the Prospector scale for the 'Empowerment'-variable may also be partly explained by an additional analysis based on company size. In this analysis a statistically significant difference *(p=0.002 for Reactor)* was found, suggesting that respondents from smaller operators agreed comparatively more to the statement: "*If an issue is identified as priority, senior management becomes involved and takes over*". Such would imply that the smaller the operator, the higher the chance that management will interfere in the execution of tasks of lower level employees if they find that this task becomes a priority issue.

9.5.5   *Strategic external organisation actions*

In Chapter 4, five variables were identified to measure the strategic external organisation action: 'environment', 'suppliers', 'benchmarking', 'regulator' and 'partnerships'. The aggregated data for the scores on all these five variables are depicted in the below scatter box diagram.

Scatter Box 11 Aggregated Strategic External Organisation Actions

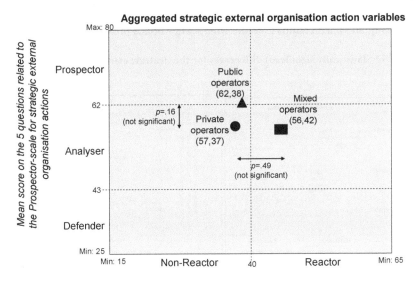

The relative positions of the WSS operators in the Scatter box suggests that the three types of companies have a very similar strategic stance towards the external organisation, All companies are Analysers and score relative moderate on the Reactor-scale. Investigating the internal consistency of the two scale by calculating the Cronbach Alpha values, a value of 0.16 (unacceptable) was calculated for the Prospector scale, while 0.54 was calculated for the Reactor scale (poor).

However, investigating the individual variables several statistically significant differences are found. For four out of the five variables significant differences are found. Only for the 'Regulator' variable no significant difference was found between the three ownership type groups (see Scatter box 12).

Scatter Box 12 Insignificant difference for the 'regulator'-variable of the strategic external organisation action

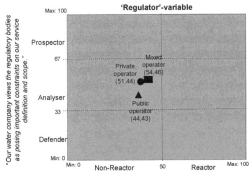

The Scatter Box suggests a similar behaviour as was found for the aggregated external organisation strategies, e.g. Analyser behaviour. For four variables a statistical difference is found related to the ownership variable: environment, suppliers, benchmarking and partnerships (see the four scatter boxes presented below).

Scatter Box 13 Statistically significant differences for the strategic external organisation variables

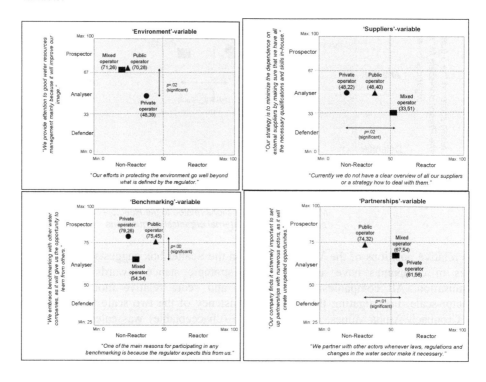

The statistical differences for all these variables go hand-in-hand with statistical differences in the regulatory environment. For none of the variables significant differences were found according to size.

For the variable 'Environment' a statistically significant difference was found on the Prospector scale ($p=0.019$) (see top left scatter diagram). Respondents from the (English and Welsh) private companies agreed least to the statement "*We provide attention to good water resources management mainly because it will improve our image*". The data indicates a Defender attitude of private (English and Welsh) companies for this variable. This can be explained by the primary profit motive of the private companies. For them there is no direct benefit to spend money on good water resources management, apart from the more long-term benefit of a better water quality to abstract. For publicly owned operators this may be different. They also have a responsibility to protect the public interest, and as a water company they may feel they are best placed to interfere also in general water management. Italian companies have in that respect a peculiar role since they are organised on catchment level scale since the Galli law. In this view it is quite logical they will be involved in integrated river basin management.

For the 'Suppliers' variable a statistically significant difference was found on the Reactor-scale (p=0.014) (top right scatter box diagram). Respondents from the (English and Welsh) private companies disagreed most to the statement "*Currently we do not have a clear overview of all our suppliers or a strategy how to deal with them*". The data suggest that private suppliers are comparatively more consistent in their dealings with suppliers, since they put effort in having a proper overview and a strategy how to deal with them. An explanation of the Non-Reactor like behaviour of private companies towards suppliers may be two-fold. On one hand, the relation with suppliers is important to them as it directly impacts their cost levels. On the other hand, the relation with suppliers may be of higher interest for private companies, especially in case they make use of daughter companies as suppliers.

For the 'Benchmarking' variable a statistically significant difference was found on the Prospector scale (p=0.002) (lower left scatter box diagram). Respondents from mixed (Italian) companies agreed least to the statement "*We embrace benchmarking with other water undertakers as it will give us the opportunity to learn from others*". Such seems logical, as the benchmarking schemes in both the UK and the Netherlands have matured in the last decade(s), while Italian benchmarking is still in its inception phase.

For the 'Partnership' variable a statistically significant difference was found on the Reactor scale (p=0,013) (lower right scatter box diagram). Respondents from public companies disagreed to the statement "*We partner with other actors whenever laws, regulation and changes in the sector make it necessary*". This would suggest that publicly owned companies are more independent from external developments in strategizing their partnership network.

9.6   Synthesis of the Chapter

This Chapter aimed the extent to which WSS operators in different institutional settings take different strategic actions. Especially of interest in this comparison is whether private operators take different strategic actions compared to public operators.

The Chapter builds firmly on the research design developed in Chapters 4 and 5. Several choices were made in Chapter 4 that have consequences for the research at hand. The first choice is to use strategic typologies to identify the strategic actions of WSS providers. Then based on investigation of alternative strategic typologies proposed by several researchers, the typology of Miles and Snow was selected, which dates from 1978. Using more recent insights on the use of the Miles and Snow typology, a two-scale framework is established. In this framework one scale measures the extent to which WSS provider may have Reactor-type of characteristics; the other scale measuring the extent to which they have Defender versus Prospector characteristics. The next important choice made in the Chapter 4 is to use a survey as the research instrument. For the survey a questionnaire was developed, including several pre-testing rounds, consisting of 44 questions; half of them assessing the Reactor scale, the other half assessing the Defender-Prospector scale. Moreover, the questions were grouped along the five strategic dimensions (markets, product/services, seeking revenues, internal organisation, external organisation). A target population was selected of WSS providers in England & Wales, Northern Ireland, Scotland, The Netherlands and Italy. The sample of WSS providers provides fully privately owned companies, mixed owned companies, and fully public companies. Of this sample in total 191 respondents were approached distributed over 64 WSS providers.

The methodology to conduct the survey was based on the Tailored Design Method of Dillman making at five occasions contact with the target population to maximize the return rate. From the selected group of respondents 96 useable responses were collected, distributed over 49 companies (e.g. a collection rate of 50% at respondent level, or 77% at company level). The first step in the data analysis was to establish whether it was justified to use the average for one company, if from this company multiple responses were collected. Based on an inter-rater analysis it showed that such was justified. Then for each of the strategic dimensions an ANOVA analysis was made whether private companies pursue different strategies compared to companies with lower levels of private investment. Overall, one could observe that ownership explains for one of the two continuums a difference in strategies for 14 out of the 22 variables.

The first section of data analysis of the survey tried to establish whether the scale of 22 indicators used to assess the strategic typology matches the overall impression of the strategic archetype of the company. Each respondent was asked to select from 4 profiles the one they found most similar to their own company. From the data analysis it shows that for approximately half of the companies the result of multi-indicator analysis was similar to the self-description.

Then for the multi-items scores, the data analysis showed that for the strategic market action the aggregated score for all four market strategy-variables gives a significant difference. Mixed companies are, according to the aggregated scores the least Reactor-like, followed by fully privately owned companies, while fully public companies act most Reactor-like with respect to the market strategies. Moreover, going more in detail, three of the composing market-variables give a statistical significant difference. Fully private companies are more Reactor-like with respect to inset appoints. Fully public companies have a Reactor-like behaviour with respect to their connection strategies, and the more private a company is, the more Defender like its strategic actions with respect to mergers and acquisitions.

For the strategic products/service actions the data showed that the more private the operator, the more it will act Reactor-like with respect to the quality of water. This is quite an interesting result. It shows that the more public, the more the WSS provider becomes pro-active in ensuring and upgrading the quality levels, while the ambition of the private companies is merely to comply with the current legislation.

For the third component of strategic seeking revenues actions several statistically significant differences were found. The aggregated indicator for all 'seeking revenues' variables shows that public companies are more Prospector-like. This can largely be explained by the scoring on one indictor, e.g. the willingness to provide water-for-free to boost its public image. In view of the more pronounced public character of publicly owned companies it seems logical they put more emphasis on serving the public cause. The data show that private companies allocate their profits in a different way compared to public companies. Fully privatised companies spent relatively lower portions of their profit on existing infrastructure. Such may be explained by the notion that their profit will go to their shareholders, but the low score of the mixed companies makes such statement questionable. For the efficiency-variable, the data showed that the more public the provider the more it acts Reactor-like.

For the strategic internal organisation action also a statistical significant difference was found for the aggregated indicator. Overall, the data shows that mixed companies are more Reactor-like compared to the other two types of companies. This may have to do with the influences both the public and the private shareholders want to exercise on the day-to-day management of the company. Especially this influence of 'outsider' surfaces in the selection of their managers and the variability in the group of strategy makers, given the significant higher scoring of mixed companies on these two indicators. For one other indicator a statistical difference was found. Mixed companies have much more a Defender attitude with respect to empowering their employees.

The last of the strategic dimensions involves the external organisation. Again many statistical significant differences were found. On the Defender-Prospector scale, fully private companies score low in taking care of the environment; and mixed companies score low on the willingness to learn from benchmarking. On the Reactor-scale, mixed companies are much less Reactor-like compared to the other two types in their dealings with suppliers; and public companies score are much less Reactor-like in engaging in partnerships.

Overall conclusion of the survey is that ownership -to a certain extent- influences the strategies pursued by WSS operators, although the influence is limited to a sub-set of the strategic actions. However, interpretation of the conclusions in terms of universality and methodology should be done with caution. Since this study focused on a population of relatively mature organizations from industrialized nations, the generalizeability of the findings may be questioned, even though these particular characteristics are eminently suitable for carrying out a study on strategy. Though there is no a priori reason to assume uniqueness of this particular population, repetition of the study in more generalized populations is called for.

In terms of methodology the main limitation is that other factors may also influence the logical sequence of 'reform measures – change in organizational attributes – change in strategy. For example, privatisation does not only entail a shift in ownership, but in many cases it also includes a shift in subsidy structures and financing mechanisms. When the English and Welsh water operators were privatised they received a substantial grant for a smooth change. Such change in financing mechanisms may influence also the strategies and the levels of performance of the operators. Hence, the logical sequence is not a straightforward causal relation and conclusions about the validity that a different ownership type has indeed an impact on the strategies of operators, with consequent changes in performance levels should be interpreted with caution.

Section III. Analysis

Chapter 10 Performance, Institutions and Strategic Actions

This Chapter addresses the analysis of performance with respect to the institutional context and the strategic actions. An analysis is carried out of to relate the performances of operators to different strategic archetypes.

10.1   Introduction

This Chapter aims at including performance in the analysis of institutional changes. In order to fulfil this objective, the analysis provided in this chapter identifies to relate the score on the variables of the strategic typologies to performance indicators. This relation was labelled in the Analytical Framework in Chapter 5 of the Research Design section as relation (3).

In the previous Chapter the scores were assessed of individual operators on the scales Defender-Prospector and Non Reactor-Reactor. These scores are in this section compared with available benchmarking data. Two relevant hypotheses are tested for the sample population.

1. The higher the score of the WSS provider on the Reactor scale, the lower the performance of the operator. The Reactor stance is associated in literature with low performance (see Chapter 4). This hypothesis is tested in this Chapter for the sample population of WSS providers.

2. The higher the score of the WSS provider on the Defender-Prospector-Prospector scale, the lower the performance. As the WSS sector composes a mature, non-turbulent environment a Defender stance might be most effective.

As can be deducted from the analysis of the OfWat International Comparator reports in Chapter 3, an international comparison of benchmarking data is highly complicated. Hence, a choice was made in this thesis to analyse the results from the survey on strategy typology with country based benchmarking results. In this respect two benchmarking data sets are relevant for the thesis, e.g. the Dutch benchmarking and the English and Welsh benchmarking.

The following two sections will relate the results from the survey with the benchmarking results for these two data sets.

10.2   Strategy and performance for the English and Welsh WSS providers

In the UK, an important driver for the comparison of performances between water companies was the privatisation of the service delivery and the starting of regulatory capacities. Since the privatisation of water and wastewater systems in the UK, the Office of Water (OfWat) has required the private companies to maintain their extensive infrastructure of water mains and sewers in a manner that provides "adequate" services to current and future customers. The goal is to ensure that private companies provide adequate investments in infrastructure while maintaining competitive rates. For the 5-year license renewal performance assessment, OfWat reviews each company's performance indicators for the previous 5 years, as well as its plans for the future O&M and rehabilitation of its infrastructure. By examining the trends over several years, OfWat determines whether the O&M and rehabilitation carried out by the company has resulted in improving, stable or deteriorating services to customers. Based on this review, OfWat in effect approves each company's capital reinvestment budget. Specifically, it approves a water company's request to set future prices at a level that will provide sufficient funds to maintain its network. Companies are required to carry out any work needed to rectify deteriorating serviceability to customers, either before license transfers or as part of the new license, but at no cost to customers. The need for such work at a license transfer would be reflected in the company value at transfer. Such a potential liability should provide an incentive for the companies to ensure that they maintain the serviceability of the water main and sewer networks.

OfWat conducts an annual review of company performance against various predefined Performance Indicators. The companies included are all the major water companies and water and sewerage companies in England & Wales[56]. These –so called- 'Levels of Service' reports include information from the Environment Agency and the Drinking Water Inspectorate (DWI) on companies' performance in safeguarding the environment and providing good quality drinking water. Next, OfWat also publishes annually financial performance and expenditure reports of the water companies in England & Wales. These reports provide insight into the relative efficiency of the water companies.

10.2.1   The strategy scores, overall performance and the individual key performance indicators

The OfWat assessment of company performance focuses on the delivery of services to customers. There are five key categories of performance indicators for assessing performance related to drinking water services: water supply, water distribution, customer service, environmental impact, and infrastructure cost. In each of these areas, OfWat has developed specific output performance on the basis that services measured should be of real importance to customers. The methodology used by OfWat consists of several indicators to assess the performance of the operators in the field of drinking water services provision, e.g.:

1. DG2: Properties at risk of low pressure. This indicator shows the number of connected properties that have received, and are likely to receive, pressure below the reference level when demand for water is not abnormal.
2. DG3: Properties subject to unplanned supply interruption of 12 hours or more
3. DG4: Properties subject to hosepipe bans at any time during the year
4. DG6: Billing contacts not responded to (within five working days)
5. DG7: Written complaints not responded to (within 10 working days)
6. DG8: Bills not based on meter readings

---

[56] Albion Water and Cholderton & District water company Ltd. are excluded from these reviews given their small sizes.

7. DG9: Received telephone calls not answered within 30 seconds
8. DG9: Telephone call handling (calls abandoned, all lines busy, call handling satisfaction).

OfWat combines the scores of the above mentioned indicators in one overall indicator, called OPA (Overall Performance Assessment), which compares the companies' performance over a broad range of measures. The OPA assessment enables OfWat to compare the quality of the overall service companies provide to customers, and to take this into account at each price review. Moreover, the OPA informs customers (and other interested parties) about the overall performance of their local water company. OfWat's methodology in calculating the OPA score is relatively straightforward. First OfWat converts each performance measure to a score out of 50 points. The better a company's performance, the higher it scores. These individual OPA scores are then weighted (to reflect the importance of that element in the total OPA score) and then added together to form the total OPA score. The weighing factors are outlined in Annex 9.

To adequately compare the performance of the WSS providers with strategies, the average OPA score is taken from multiple years. In the below graph the scores over six years are represented for the participating WSS providers in the strategy survey.

Graph 5 Overall Performance Assessment (OPA) from 2001 to 2007

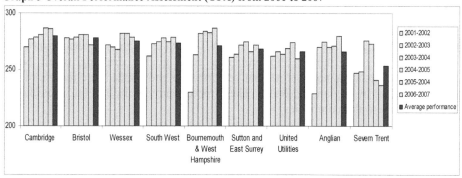

It is noted that none of the indicators included in OPA relate to an evaluation of the efficiency of the services provided. Hence, apart from the indicators included in OPA, also efficiency indicators were selected from the financial annual reports published by OfWat. The Strategy- Performance relationship was tested by calculating the Pearson r value[57]. Table 33 depicts the relations expressed in the Person Product-Moment coefficient between the strategy scores and the English OPA scores:

---

[57] The Pearson Product-moment coefficient is calculated to analyse the extent to which the scores obtained in the strategic typology survey proof similarity with the performance ratios. The Pearson test identifies whether two series of data show similarity; not the causality. If the correlation coefficient (indicated as r) would be 1.00 or -1.00 (inverse) there would be a complete identical ranking of the strategies and the OPA performances; while an r-value of 0.00 shows a complete lack of similarity. If a correlation coefficient is significant this means that the correlated series of data show more similarity than chance would expect. Often significance is being checked by squaring the value of r. The result is a percentage between 0% and 100%.

Table 33 Strategy scores and the English OPA scores

Calculation of Pearson's r			REACTOR SCALE						DEFENDER - PROSPECTOR SCALE					
			Combined Reactor	Market	Products/Services	Seeking Revenues	Internal Organisation	External Organisation	Combined Prospector	Market	Products/Services	Seeking Revenue	Internal Organisation	External Organisation
OPA		AVG	.1	.1	.2	.0	-.2	.6	.2	-.4	.5	.2	.1	.3
		Gr.	-.1	.1	.1	-.4	.0	-.1	.0	.0	.0	-.1	-.2	.3
Water supply	DG2: pressure	AVG	-.2	.2	-.8	.2	-.2	-.4	.2	-.1	.6	-.6	.4	.2
		Gr.	-.5	-.3	.3	-.2	-.7	.0	-.4	-.4	.2	-.7	-.1	.1
	DG3: interruptions	AVG	.2	.5	-.5	.2	.0	.6	.6	.0	.9	.3	.4	.4
		Gr.	.0	.2	.5	-.5	.1	.0	-.2	.1	-.2	.3	-.5	.0
	DG4: hosepipe bans	AVG	-.2	-.4	-.1	-.4	-.3	.4	.3	.4	.1	.1	-.2	.5
		Gr.	-.2	-.4	-.1	-.4	-.3	.4	.3	.4	.1	.1	-.2	.5
Water quality		AVG	-.4	.3	.1	-.8	-.4	-.4	-.3	.2	.0	.1	-.7	-.1
		Growth	-.2	.0	-.6	-.3	.1	-.4	.4	.8	-.2	.2	-.1	.2
Customer service	DG6: billing	AVG	.4	.2	.5	.2	.2	.4	-.1	-.5	.2	.1	.1	.1
		Gr.	.5	.0	.4	.2	.3	.8	.2	-.5	.4	.2	.3	.4
	DG7: complaints	AVG	.4	.2	.6	.2	.1	.5	-.2	-.5	.1	.2	.0	-.2
		Gr.	.2	.0	.0	.3	.0	.5	.1	-.7	.6	-.4	.5	.4
	DG8: readings	AVG	.7	.4	.3	.3	.5	.6	.1	-.5	.4	.3	.3	.1
		Gr.	-.4	-.5	.6	-.5	-.2	-.2	-.5	.1	-.8	-.3	-.6	-.2
	DG9: telephone	AVG	.4	.1	.5	.5	.2	.2	-.4	-.7	.0	.0	.2	-.5
		Gr.	.0	.2	.0	-.3	.2	-.2	-.1	.0	-.1	-.2	-.2	.0

Table 33 shows that most values are relatively low, suggesting a low correlation between the strategic typologies and the performances. There are some exceptions to this general observation.

Only one higher value correlation of at least .90 is found, namely the one between the Prospector score on products/services strategies and the performance indicator of interruptions. Graph 6 depicts the tight relation between the strategy scores and the performance with respect to interruptions.

Graph 6 The 'products/services'-strategy score and the 'interruptions'-indicator

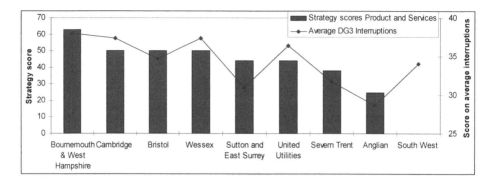

The Graph suggests that companies that have a very progressive, active and aggressive strategy with respect to diversifying their product portfolio, also score good in being able to minimize the number of interruption in their drinking water supply. Two arguments may provide the rationale for such relation. First it may be that only those companies that have their own 'house' in order, will have the willingness to investigate possibilities to diversify their portfolio. Another argument could be that in diversifying their products, companies will have the ability to materialize some economies of scope, which can be beneficial to the core operations. Such could in this case be for example, that the companies have also gained added value expertise by offering consultancy services on leakage assessment, which can also be applied 'at home'.

Four other correlations were also found with Pearson's r values higher than 0.8:
* The more the company is non-Reactor like in its products/services strategies; the higher it scores in maintaining the pressure in the network.
* The more the company is non-Reactor like in its seeking revenues strategies, the better the water quality.
* The more the company is Reactor like in its external organisation strategies, the better it deals with responses to billing contacts.
* The more the company is Defender-like in its products/services strategies, the higher growth ratios it has managed to realize in basing its bills on actual meter readings.

However, in view of the limited number of statistically significant correlations between the levels of service indicators and the strategy scores the hypothesized relation between strategies and performance is to be rejected.

10.2.2   The relation with financial performance indicators
The previous section did not find a strong relation between the OPA (and its components over six years, both in terms of the average as for the growth rate, and the strategy scores. However, instead of taking OPA, one can also look at financial performance indicators over the years.

OfWat publishes annual reports regarding financial performance and expenditures of the English and Welsh water providers. The information provided in these reports complement the 'Levels of Service' reports discussed earlier. One of the prime financial performance indicators is the Return on Capital Employed (ROCE). ROCE is commonly used as a measure for comparing the performance between businesses and for assessing whether a business generates enough returns to pay for its cost of capital. In the below Figure the ROCE's achieved by the individual companies are presented.

Graph 7 Return on Capital Employed of English & Welsh water providers

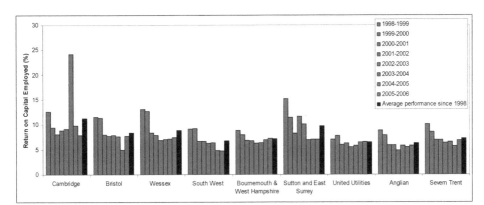

Both the average ROCE, as the ROCE growth rate over the last 8 years is selected to relate them to the strategy scores. In addition to ROCE, OfWat makes since last year a ranking of the companies with respect to its operating efficiency, and its capital maintenance efficiency. Also these rankings are used to compare them with the strategy scores. The comparison results in the following Table, indicating Pearson's r for each relation.

Table 34 Financial performance and strategy scores in England & Wales

| Calculation of Pearson's R | REACTOR SCALE | | | | | | DEFENDER - PROSPECTOR SCALE | | | | | |
	Combined Reactor	Market	Products/Services	Seeking Revenues	Internal Organisation	External Organisation	Combined Prospector	Market	Products/Services	Seeking Revenues	Internal Organisation	External Organisation
Best ranking operating efficiency (low is good)	.0	-.6	-.3	.5	.0	.0	.2	.0	.0	-.2	.5	.2
Best ranking capital maintenance efficiency (low is good)	.9	.1	.3	.6	.9	.6	.2	-.2	-.2	.6	.4	-.1
Return on Capital employed (%) AVG	-.2	.3	.0	-.1	-.4	-.2	-.2	-.2	.3	.0	-.2	-.2
Return on Capital employed (%) Growth	-.2	.5	.1	.0	-.4	-.2	-.4	-.3	.5	-.4	-.1	-.4

Based on the Table it can be concluded that there is no strong relation between the strategies and the performance of water providers in England & Wales. There is one notable exception. The data suggests that companies with low Reactor scores perform better in efficient capital maintenance.

Graph 8 The Reactor-strategy score and capital maintenance efficiency

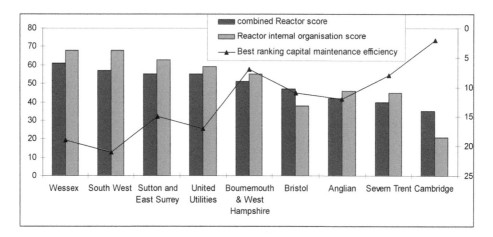

The rationale for companies with a low Reactor score to have a better capital maintenance performance may be explained by the following interpretation. It may be that a company is able to realize lower capital maintenance costs because it restrains itself from reacting to short-term expensive desires and demands from the regulator. Instead such company will fare its own course and has the ability to economize due to its long and steady planning horizon.

10.3   Strategy and performance of Dutch WSS providers

In the Netherlands, since 1989, an informal association of a number of water supply companies named COCLUWA has been involved in metrics benchmarking of its member companies. The COCLUWA initiative evolved into the VEWIN voluntary benchmarking study named 'Reflections on Performance'. The study was aimed at providing insight into the performance of water providers and to formulate best practices to enable them to further improve business processes and performance. Since the start of the VEWIN benchmarking four reports have been published at a three year interval. Starting in 1997 with 18 water providers (equivalent to 85% of the Dutch drinking water sector), the latest report from 2007 included 10 Dutch water providers, covering almost 100% of the total number of connections in the Netherlands. Industrial water and non-drinking water activities fall outside of the scope of the benchmark exercise.   In the plans for the revision of the 'Waterleidingwet' a compulsory benchmark is incorporated. The aim of the benchmarking exercise was to provide the industry with insight as to how water companies might improve their processes, and to better explain things to interested parties including commissioners and shareholders (Blokland *et al.,* 1999).

The indicators used in the program include water quality, service levels, environment and finance and efficiency (see Table below):

Table 35 Key performance indicators and results of the VEWIN benchmark

Main area	Performance indicator	Conclusion 2007 report
Water quality	Water Quality Index	The benchmarking results show that all companies are very close to the optimum score with only a minor variation between the companies; with the best companies reaching 0.02, while the lowest performing companies reach 0.04.
	The number of times they exceeded the legal standards.	Scores vary between 0.021 and 0.170.
	Perceptions on the quality of the water	The best companies score for this benchmark indicator an 8.4. The least performing score a 7.4.
Service	General satisfaction	Best performing companies score a 7.9, least performing score a 7.4
	Percentage of calls answered within 20 seconds	Best performing company manages to answer 79% of the incoming phone calls within 20 seconds, while the least performing succeeds in only 17%.
Environment	Use of energy Produced residues Land hydration Contributions to nature management	No scores are provided of the individual water providers in the reports.
Finance and Efficiency	Costs per connection	The most expensive has Euro 245 per connection, while the lowest has Euro 167 per connection.
	Operational costs per connection	The most expensive has Euro 134 per connection, while the lowest has Euro 76 per connection.
	Costs per m^3	The most expensive has a cost/m^3 of Euro 1.72, while the cheapest delivers water for Euro 1.06/m^3.
	Operational costs per m^3	The most expensive has Euro 0.96/m^3, while the lowest has operational costs of Euro 0.44/m^3

Unlike the OfWat benchmarking no overall performance assessment indicator is generated. Hence, to be able to make a comparison with the scores obtained of the strategies, an effort is to be made to combine the above mentioned indicators into one ordinal scale.

In order to do so, several choices are made to make the comparison. First, only performance indicators are used of which data is available in the benchmarking reports of individual companies. Hence, a choice is made to use 9 performance indicators. The second choice is to give equal weighting to each of the performance indicators. A third choice is to only use the data from the two latest reports from VEWIN (from 2003 and 2006). Although it would be better to use the average performance of the companies over a longer period, due to the constant mergers in the Netherlands it is not possible to make a sound comparison on a longer period of data.

The procedure to compare the data is by establishing an index value from 1 to 100 for every performance indicator. Hence, the maximum any water company can score is 800 points. The sum of all the performance indicators is used as an indicator to compare with the strategy scores. The below Graph 9 presents the scores of the water companies expressed by their index values.

Graph 9 Index scores of Dutch water providers

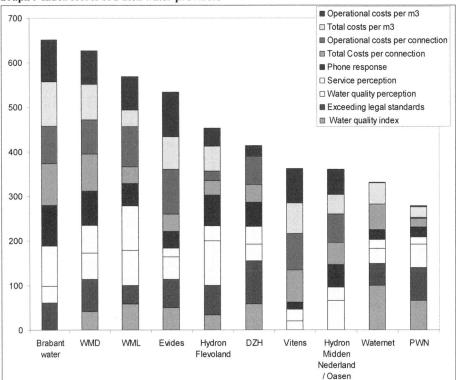

Pearson Product-moment coefficient is calculated to analyse to which extent the scores obtained in the strategy survey proof similarity with the performance ratios. The results are presented in Table 36. In the Table no correlation coefficient is found of at least .90. Hence, no relation can be established between the strategic actions and the consolidated performance of the Dutch WSS providers.

Table 36 Strategy scores and the Dutch performance indicators

Calculation of Pearson's R		REACTOR SCALE						DEFENDER-PROSPECTOR SCALE					
		Combined Reactor	Market	Products/Services	Seeking Revenues	Internal Organisation	External Organisation	Combined Prospector	Market	Products/Services	Seeking Revenues	Internal Organisation	External Organisation
Dutch OPA	Avg.	-.1	-.3	-.3	-.0	-.3	.4	.4	.5	.3	.0	-.0	-.4
Water quality	Avg.	.2	.7	-.1	.0	.4	-.5	-.5	-.6	-.1	-.5	-.1	.5
	Grow.	-.1	-.6	.1	.1	-.5	**.8**	.2	.4	.0	.4	.0	-.5
Norm violations	Avg.	-.1	-.1	.7	.0	-.3	-.2	.2	.0	-.4	.0	.1	.1
	Grow.	-.1	-.1	.8	.0	-.3	-.2	.1	-.2	-.5	.0	.1	.3
Water quality perception	Avg.	-.2	-.2	-.4	-.1	.0	.1	-.2	-.1	.2	.6	-.1	-.5
	Grow.	.5	.1	.3	.6	-.2	.4	-.1	.0	-.2	-.4	-.4	.1
Milieu belasting index	Avg.	.1	.7	-.1	.2	-.4	.0	-.5	-.6	.0	-.2	-.3	.3
Service perception	Avg.	-.3	-.1	-.1	-.2	-.5	.3	.2	.1	.2	-.2	-.1	.2
	Grow.	.0	.2	.0	.3	-.4	.0	-.1	.2	.1	.1	-.3	-.4
Phone response	Avg.	.0	-.2	-.4	.0	.1	.1	.2	.4	.3	-.1	-.2	-.1
	Grow.	.5	.0	.1	.5	.1	.3	.3	.5	.1	-.7	-.4	.2
Costs / connect.	Avg.	.2	.3	-.3	.1	-.2	.5	.0	-.1	.4	-.1	.3	-.2
	Grow.	.5	.2	.1	.5	.1	.1	.1	.3	.1	-.4	-.2	-.4
Oper. costs / connect.	Avg.	.2	.2	.1	.2	.0	.0	-.6	**-.8**	-.4	-.1	-.3	.4
	Grow.	.6	.1	.2	.7	.1	.3	.1	.1	-.1	-.5	-.2	-.2
Costs/ m³	Avg.	.4	.3	-.1	.3	.1	.4	.0	-.1	.3	-.3	.1	.3
	Grow.	.5	.1	.2	.5	.0	.2	.2	.3	.0	-.4	-.2	-.2
Oper. costs/m³	Avg.	.4	.2	.1	.3	.2	.1	-.4	-.6	-.3	-.2	-.2	.5
	Grow.	.6	.1	.3	.7	.0	.3	.1	.1	-.2	-.5	-.3	-.1

Hence, no correlation is found between the strategy scores and performance scores in the Dutch WSS sector. Only in two instances there is a fair degree of similarity between the strategy scores and the performance scores. One is that organisations that score well in the water quality index are also very Reactor-like in their external organisation.. The other is that organisations with low operational costs are also very Prospector-like in their market actions.

10.4   Synthesis of the Chapter

By analysing the relation between performance and strategic actions, the Chapter includes a new element to the research. In the earlier Chapters, the analysis focussed on all the factors that may change the performance, but in this Chapter it is verified whether such is indeed the case.

The Chapter clearly shows that to interpret measure and compare the performance of WSS providers several fundamental difficulties occur. These problems are more important for non-profit organisations, like the majority of WSS providers. WSS providers depend on the perceptions to interpret how well it is operating, making its performance locally dependent and serving a multitude of constituencies. However, it can be noted that the measurement of performance and the identification of performance indicators has taken large proportions in the WSS sector. Such is mainly undertaken in many benchmarking exercises around the world.

The presented study tested the relation between strategic actions and performance. Although outside of the WSS sector, this relation has been analysed earlier by many researchers. An overview is presented of a selection of these studies. Based on the evidence provided in these studies any simple relationship should be rejected. Variables both of external and internal nature add complexity to the relationship, like size, external environment, executive characteristics, technological deployment, knowledge orientation, and the strategy formulation process. However, based on the overview two hypotheses were formulated to test for the target population, e.g. the performance of a WSS provider would be higher if it has a relatively more Reactor, and Defender characteristics. These hypotheses are tested for the target population, using as input the outcomes from the strategic actions assessment from the previous Chapter, and benchmarking data from the various countries. In England & Wales the benchmarking data is sourced from OfWat. OfWat has composed an Overall Performance Assessment (OPA) indicator, which is aggregated from a basket of indicators. Correlation the ranking of the average OPA of the English & Wales WSS providers with the strategic typologies only one statistical significant correlation coefficient was found. It was found that if an English & Welsh WSS provider has more Reactor-like characteristics with respect to its strategic internal organisation actions, its OPA score will be higher. In the Netherlands the benchmarking data provided by VEWIN was used. Since VEWIN does not compute a similar aggregated indicator like the OPA indicator of OfWat, one was separately calculated. However, for the Dutch OPA scores no statistical significant correlation was found with the strategy scores from the survey.

Hence, conclusion is that the hypothesis of a strong relation between strategic actions and performance should be rejected. From the studies no clear evidence is found that a different strategic typology is conducive to higher (or lower) performance levels. The inherent problems with performance interpretation, measurement and comparison form a prime obstacle for researchers to test the relationship between institutional arrangement, strategies and performance. This conclusion has a two-fold effect on the thesis. On the one hand it is supporting the relevancy of the thesis as it underlines the importance to analyse whether institutional changes may affect the conduct of WSS providers (since it is difficult to relate them directly with performance). On the other hand, it weakens the basis of the thesis as it becomes unclear whether it is useful to focus on strategies to analyse the change in conduct. Apparently, it can not be stated for sure that a change in strategies has an effect on performance.

Section IV: Synthesis

Chapter 11:   Synthesis of the Thesis

Section IV. Synthesis

Chapter 11 Synthesis of the Thesis

This Chapter draws conclusions from the previous 10 Chapters. It identifies numerous issues that may require further discussion. Finally, it sets out possible directions for future research.

11.1   Introduction

The aim of the thesis is to provide further insight into the neo-liberal reforms in the WSSS sector. In this respect a total of 10 Chapters has been presented composed of theoretical notions, case studies, survey material and presentations of intermediate research results. Now, this final Chapter will synthesise and draw conclusions from the material presented. It will also identify elements that arise from the thesis which are subject to discussion, and   elements which may proof valuable for further research.

11.2   The research relations reviewed

The two Chapters making up the Research Design section establish the testable analytical path model to achieve the research objective. Interpreting "strategy", a distinction is made in 5 layers of analysis, e.g.:

1. Does neo-liberalism make a difference for the institutions that govern the WSS sector?
2. Do neo-liberal institutional changes make a difference for the strategic context of WSS providers?
3. Do neo-liberal institutional changes make a difference for the strategic plans of WSS providers?
4. Do neo-liberal institutional changes make a difference for the strategic actions of WSS providers?
5. Do different strategic actions make a difference for the performance of WSS providers?

In the Analysis section analysed each of the five relations is addressed in five respective Chapters. The main conclusions with respect to the 5 relations are presented in the below completed Analytical framework.

Figure 19 Analytical framework of the thesis – completed

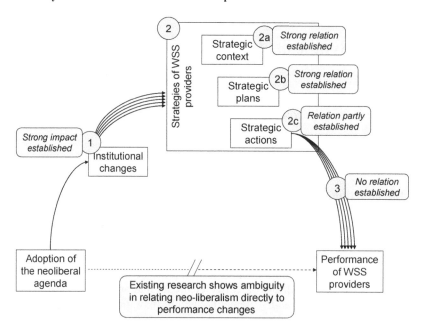

The analytical framework gives the structure to the Analysis Section of the thesis. The following sections will address the conclusions for each of the 5 relations.

11.2.1   Institutional dynamics (1)

Institutions are interpreted in the thesis through shifts in the relation between the responsibility entity (the government) and the management entity (the WSS provider). This is investigated through a case study of the Western European WSS sector. The case study showed the large impact neo-liberal changes had, have and will have on the institutions of the WSS sector. Many Western European policy makers have attempted to involve private parties and make use of the market mechanism in the WSS sector. The case study clearly showed that further proliferation of the neo-liberal agenda in the Western European WSS sector is dependent on numerous drivers and constraints. It was found that any driver for further adoption of the neo-liberal agenda in one country can be a constraint in another country. Next, the analysis showed that the institutional landscape of the WSS sector is at drift with a clear move towards more delegation, and more private sector involvement. A strong trend was identified throughout Western Europe from Direct Public Management, to Delegated Public Management, to Delegated Private Management. Most countries in Western Europe find themselves in a transition phase from one institutional arrangement to another. However, the vast majority is reluctant in adopting the most extreme manifestation of neo-liberalism, i.e. full divestiture. Only one country has gone all the way with adopting the neo-liberal agenda through Direct Private Management, e.g. England and Wales. The scenario analysis taught us a clear path dependency of the development of the Western European institutional landscape. The identified drivers for change and the scenario analysis showed that it is plausible that the trend of delegation and private sector involvement will continue, and policy makers will continue to explore alternative institutions to increase the performance of the sector. As Melosi (2004; 221) stated:

> for a very long time municipal control or oversight of water service has been part of the local fabric of cities - a venture that set precedents for many other services to follow. The new century may have something different in store for municipal water supplies and delivery as fresh water becomes an increasingly scarce resource and a more profitable commodity. The line between water as a public good and water as a product always has been blurred, but never to the extent it is today.

The result from this first level of analysis is important as it confirms the relevancy of the thesis, although the outcome is neither surprising, nor innovative. Policy makers have a direct influence on many of the institutions that govern the sector, especially the regulatory framework. It is quite obvious that institutions change if policy makers adopt the neo-liberal reform agenda. What is striking, however, is the magnitude of the past changes and the plausible future development. The analysis shows that almost no country in Western Europe has been isolated from the neo-liberal changes. Moreover, the analysis of the drivers for change and the scenario analysis show that it is plausible that involvement of private parties and introduction of competition will remain high on the agenda of policy makers contemplating on improving the sector's performance.

The next three Chapters 7, 8 and 9 aimed to establish whether different neo-liberal institutions changes are conducive to different strategies of WSS providers. To understand this relation a distinction was made based on a three level analysis, e.g. the relation of institutions with strategic context (2a), with strategic plans (2b), and with strategic actions (2c) of WSS providers.

11.2.2   The relation between institutions and strategic context (2a)

This layer of analysis is concerned with whether WSS providers in different institutional contexts *can* have different strategies. The relevant hypothesis for the relation between institutional change and strategic context was that institutions make a difference for the strategic context of the WSS provider. Institutional changes are interpreted through the regulatory regimes of the WSS provider, while strategic context is interpreted through the managerial discretion of the WSS provider. The research question was investigated through a comparative case study of England & Wales (neo-liberal) and the Netherlands (traditional public).

The insight gained from the case study support the hypothesis, in particular for the strategies related to markets, products and services, and seeking revenues. For the strategies related to the internal and external organisation the hypothesis is to be rejected since no major difference was found for these strategic components. It was found that both the set up of the regulatory regime and the effect it had on the managerial discretion was quite different in both countries. England & Wales relied primarily on centralized national regulators, like OfWat and the Environment Agency; while the Netherlands relied mostly on decentralised and scattered regulation through company bylaws, provincial directives, instructions from water boards, or national legislation.

From the analysis it came that the differences between the impositions and opportunities provided by the two contrasted regulatory regimes found their origin in the difference in the perception of the entitlement of individual customers. The English and Welsh conceptualise that each individual domestic and industrial customer is entitled to receive WSS services from any provider, irrespective of where they live, while the Dutch find that only a portion of the customers (the 'footloose' customers) have this ability. As such, the English and Welsh WSS operators have relatively more discretion with respect to market strategies, which is also found for the products and services strategies. The larger discretion enjoyed by the English WSS provider for market and products and services strategies is corrected through severe impositions with respect to the 'seeking revenues' strategies. For the internal and external organisation strategies, only minor differences were found.

The case shows that neo-liberal institutional changes influence what a manager of a WSS provider *can* do. For the thesis this is an important finding as it gives a first inclination that neo-liberal institutional changes also influence what managers of WSS utility will do; and ultimately whether it also affects the outcomes (performance) of such different conduct.

11.2.3   The relation between institutions and strategic plans (2b)

The relevant hypothesis for the relation between institutional change and strategic plans was that institutions make a difference for the strategic plans of the WSS provider. In this respect the central element of the element is to find out whether providers *want to* have different strategies. In this layer, institutional changes are interpreted through differences in the ownership structure, while strategic plans are interpreted through bidding documents. The analysis conducted was based on a single case study located in St. Maarten. In the case study the strategic plans of both a public operator and a private operator for the WSS provision for the coming 10 years were compared, along the five strategic dimensions of Boyne and Walker. The outcome of the analysis was that indeed ownership matters for several strategic planning components.

The analysis showed that the private party is more innovative, opportunistic and risk taking compared to the public party. The private provider intended to make the largest investments, with a rapid connection policy, with the lowest unaccounted for water rates, with the cheapest external financing, and the largest challenges towards its external organisation. For some strategies the intentions of both parties were similar, but in these cases the similarity may be explained by the impositions posed on the parties. For example, both parties asked for the same tariffs, but such was on instruction of the government; both parties proposed the same type of technology, but again such may be explained by earlier recommendations to the government. Analytical generation indicates that this difference can be accounted to the ownership structure. Again this is an important conclusion for the thesis. It appears that indeed private and public operators write different strategic plans. However plans do not have an impact on performance, but actions do. Therefore a third step of analysis is needed (see the next section).

11.2.4   The relation between institutions and strategic actions (2c)

The two previous layers of analysis addressed whether operators of different (neo-liberal) nature *can* and *want to* have different strategies. Now, it is analysed whether the providers indeed *do have* different strategies. The relevant hypothesis for the relation between institutional change and strategic actions is that the institutional context makes a difference for the strategic actions of the WSS provider. Institutions are interpreted through the ownership variable, and a distinction is made between fully private, mixed, and fully public WSS providers. Strategic actions are interpreted through four different strategic typologies, e.g. Prospector, Analyser, Defender and Reactor. Based on a survey, this hypothesis can only partly be accepted. For many strategic actions, it does not matter in which institutional context the WSS provider is situated, but for some it does.

The analysis on which this conclusion was established was based on a survey conducted under 96 respondents spread over 49 operators in (private) England & Wales, (public) Northern Ireland, (public) Scotland, The (public) Netherlands, and (mixed and public) Italy. The strategic actions of the operators in the survey were categorized in the strategic typologies of Miles and Snow, along the five strategies of Boyne and Walker. Each of these five strategies was assessed by multiple sets of variables. For two out of the five strategies a statistically significant difference was found for the consolidated set of variables, e.g. for the 'seeking revenues' strategy on the Defender-Analyser-Prospector scale, and for the 'internal organisation strategy' on the Reactor scale.

The data from the consolidated set of variables measuring the 'seeking revenues'-strategic component showed that publicly owned operators have a more Prospector like behaviour compared to privately and mixed owned operators. Such is primarily explained by the larger portion of profit spent by public companies on purposes other than their current infrastructure; and that public companies are providing more water-for-free to strengthen their image in society.

Also the consolidated set of variables measuring the 'internal organisation'-strategic component showed a statistical significant difference on the Reactor scale. The data showed that mixed companies are more Reactor like compared to publicly and privately owned companies. Mixed companies perceive to be more dependent on external pressures in formulating and pursuing strategies. Such is explained by the statistically significant differences found for three variables. First, the selection of managers in mixed companies is more dependent on external sources; second, the members of the strategy planning processes in mixed companies are frequency changing and not very consistent. Third, the employees of mixed companies have relatively the lowest level of empowerment compared to the other types of providers.

The results from the survey show that several statistical significant differences were found with respect to the strategies of WSS providers in different institutional arrangements. For 3 out of the 5 aggregated strategic components from Boyne and Walker statistically significant differences were found:

- Public companies are more Reactor-like in their market strategies, while mixed companies are the least Reactor-like.
- Public companies are more Prospector-like in their seeking revenues strategies, while mixed are the most Defender-like.
- Mixed companies are more Reactor-like in their internal organisation strategies, while private companies are the least Reactor like.

Looking more in detail at the composing variables, it was found that out of a total of 22 strategic variables for 14 variables statistically significant differences were found.

- The more private the ownership of the provider, the more this company will have Defender characteristics with respect to its mergers and acquisition strategies, its strategies to protect the environment, and its gratuities strategies.
- The more private the ownership of the provider, the more this company will have Reactor characteristics with respect to securing the quality of its services, its efficiency strategies, and its policies towards partnerships with external parties.
- Mixed companies are more Defender-like in empowering their employees, the allocation of its profits and its attitude towards participation in benchmarking schemes. .
- Mixed companies are most Reactor-like and private companies least Reactor-like with respect to the knowledge and skills of their managers, the actors involved in formulating their strategies, and their strategies towards their suppliers.
- Mixed companies are least Reactor-like in their strategies in connecting new customers, and their attitude towards inset appointments.

The above-described findings are, in the light of the earlier analyses, an important result. Essentially these findings suggest that WSS providers operating in a different institutional arrangement write different plans and pursue subsequently different strategies. Hence, based on the analysis over the three levels, the collected evidence suggests that different institutions generate different strategies of WSS providers.

11.2.5   *The relation between strategic actions and performance (3)*

The relation between strategies and performance comprises the last level of analysis, and builds on the previous analysis in 2c. The central element is to find out whether a Prospector, an Analyser, a Defender or a Reactor stance is conducive to higher performance. The analysis relates the outcome of the survey on strategies to benchmarking data from the Netherlands and England & Wales. The relevant hypothesis is that there is indeed a relation between strategic actions and performance. Such is based on the theoretical notion that certain strategies can be associated with lower performance. Several authors in multiple industries have found that when an organisation has –in the terminology of Miles and Snow- strong Reactor and Defender-like characteristics its performance will be lower. Hence, to test whether such is also the case in the water industry, a comparison was made between the degree to which a provider has Defender and Reactor like characteristics and their relative performances (as interpreted through benchmarking exercises by the Dutch association VEWIN, and the English regulator OfWat). This hypothesis was tested based on benchmarking data from England & Wales and the Netherlands. The data showed however, that such a relation could not be established. No statistically significant relation could be established between the benchmarking scores, and the scores of the providers on the Non Reactor-Reactor scale.

11.3   Conclusion

Neo-liberalism is the dominant paradigm in the WSS sector. Clearly it is relevant to gain additional insight into the value of neo-liberal institutional changes in the WSS sector. The relation between neo-liberal institutional changes and performance has gained considerable attention of scholars and a relatively large body of empirical enquiries is available. From the overview of existing investigations it becomes clear that no convincing evidence exists that either the private or the public party is superior to the other. Both the theoretical and empirical evidence of the value of neo-liberal institutional changes is ambiguous for the WSS sector.

The inability of the existing studies firmly underlines the requirement to find alternative ways in the research design. The thesis at hand intends to suggest such alternative approach by including in the analytical model the construct of 'strategy'. 'Strategy' of WSS providers is used in the research model to bridge the essentially indirect relation between institutional changes and performance of WSS providers. In this respect the thesis makes use of strategic management literature subdividing strategic actions into five components. It selects for each of these five strategic actions a set of relevant indicators. The logic of the alternative research design is that since it is not clear whether neo-liberal institutional changes have inflicted on performance of WSS providers, it may be valuable to know if these changes at least make a difference for the conduct of the WSS providers.

The research objective is only partly achieved with respect to the value of neo-liberal institutional changes. The study does show that WSS operators in different institutional context strategize differently, but these different strategies cannot be linked to difference performance levels. WSS providers operating in a neo-liberal institutional context *can*, *want to*, and -to a certain extent- also *have* different strategies compared to the traditionally public WSS providers. The evidence suggests that neo-liberal institutional changes make a difference for the strategies of the operators. The analysis shows that WSS service providers operating in a neo-liberal context are less risk averse and more innovative, both in their plans as in their strategic actions. The thesis did not succeed in establishing clear evidence that the different strategies can be linked to difference performance levels, but such may be ascribed to the difficulty to interpret performance. In this respect the study at hand is just as handicapped in interpreting performance ratios as the existing body of research. There are many problems with measuring performance. The difficulty to compare performances was one of the main arguments to study the conduct (strategies) of WSS providers.

11.4   Discussion

The central aim of the thesis is to find out whether neo-liberal institutional changes influence WSS providers in terms of changing behaviour and changing performance. For policy makers the study may help to understand the implications of their neo-liberal reforms. For the scholarly community the study is valuable as it may provide new evidence using an alternative methodology, as they have themselves tried relentlessly to provide the results themselves.

Strategies of WSS providers have received up to now scarce interest of scholars. The thesis suggested an alternative manner to test the influence of the neo-liberal institutional changes. Compared to other sectors, this is new for the sector, as little has been researched on the strategies of WSS providers. The WSS sector has been isolated from insights from theoretical developments in other sectors due to its' relatively unique character as (partly) monopolistic and the nature of the water services. The identification of strategies of WSS providers through research can in itself already be regarded as a contribution to the existing body of research on the WSS sector. It is observed that two developments make strategic management literature more relevant to the WSS sector. On the one hand, one can observe a shift in strategic management literature to less competitive sectors. Increasingly it is emphasised that strategies are not merely there to beat competitors but that they should be viewed as a means to create value in an organisation. On the other hand, one can see that practitioners in the WSS sector increasingly make use of strategies in their day-to-day operations. Hence, the isolated character of the WSS sector seems to become blurred, partly due to the influence of neo-liberalism.

The relation between conduct of WSS providers and performance of WSS has not received much attention in literature. The main literature either focuses on the reform rationales or on the performances of the WSS providers. Scholars have largely neglected 'conduct' that falls within the managerial discretion of WSS providers. The specific attempt to relate different strategic typologies of WSS providers to performances creates a window for new research. Research lines in this respect could address, apart from 'strategy', numerous types of actions of WSS providers and their relation to peformance.

Most of the limitations of the research relate to the required choices in the research methodology. For example, arbitrary choices are made in distinguishing operators in a neo-liberal institutional context from operators in a traditional context through only one variable (ownership). Another limitation refers to how to interpret performance. Performance of a WSS provider is dependent on the perception of stakeholders, and in this sense a multi-dimensional, locally dependent construct. To operationalize performance for the research, the choice was made to use benchmarking indicators. Also limitations like the multi-market dimension of the WSS sector, the dependency on the data processing methodology selected, and the dynamic nature of neo-liberal institutional changes call for caution in interpreting the research outcomes.

11.5   Future research

The thesis opens a window of future research due to the novelty of both the research topic and the methodology employed. In Chapter 5 the limitations of the current research were identified. Each of these limitations can be viewed as a constraint for the thesis at hand, but also as an opportunity for further research.

The first limitation identified, refers to the choice for a cross-sectional approach in the thesis. Further research using a longitudinal approach on the influence neo-liberal changes exercise on the conduct of WSS providers will surely provide an important addition to the existing body of research. Such studies will enrich the inclination of the study at hand that institutional changes make a difference for the conduct of WSS providers.

The second limitation refers to the selection of 'strategy' to proxy the conduct of WSS providers. Since in the study at hand the novelty lies primarily in the application of concepts from strategic management to the WSS sector, this creates also a severe limitation to interpreting the research outcomes. Tested strategic analytical models of WSS providers were lacking to base the research in. Hence, a choice was made to use relatively long-lasting theoretical models from strategy literature. Most notably the Miles and Snow typology from 1978 features prominently in the thesis. Of more recent nature, theoretical constructs from Boyne and Walker (2004) and De Wit and Meyer (1994) complement to the analytical framework. Strategies were understood along the five dimensions distinguished by Boyne and Walker into market strategies, products and services strategies, seeking revenues strategies, internal organisation strategies, and external organisation strategies. It is hoped that the application of these models to the WSS sector will be succeeded by other attempts to make the analytical framework more robust.

Although it was not a specific objective of the thesis, the material presented shows the relevance of the field of strategic management to the WSS sector. In this respect, the study encourages application of theories from other sectors. More empirical research in the WSS sector and more cross-pollination from theories from other sectors will increase the insight into many of the current issues in the WSS sector. The application of theories and insight from strategic management literature is highly relevant to the sector and becoming more prominent in the future. WSS providers will be increasingly dragged out of their isolation and become more and more like a regular enterprise. Researchers could play a role in this by adapting theories from other sectors and striving to produce conclusions that can be generalised by combining case studies with empirical research.

The evidence presented in the thesis points into the direction that institutions and conduct are related to one another, but more evidence is required to establish a stronger relation. More efforts should be taken to develop theoretical frameworks to overcome the difficulties in data comparison and consolidation. Further research may use alternatives from 'strategy' to proxy 'conduct' or by going into more detail of a strategic component (or strategic variable) identified. For example, a detailed analysis of the influence of neo-liberal institutional changes would be very interesting on highly topical subjects in the WSS sector like Unaccounted for Water, Water Demand Management, water tariffs, asset management, or financial sourcing.

Next, it may be that the inclusion of additional variables in the data processing methodology will enrich the results. Especially, refinements in local contextual factors may enhance research outcomes. In the study at hand there is strong dependency on the local country specifics. A better understanding of the influence of contextual factors will support policy makers in their weighing of alternatives.

Also further research on how to assess and to compare performance of WSS operators is valuable. The research at hand makes clear the difficulty to operationalize 'performance' of WSS providers. As performance in the WSS sector depends on the perception of stakeholders, specific research methodology should be developed how to proxy this. For example, research methodologies from non-market valuation, as used in water resources economics may be applicable. These methodologies should be used next to the existing methodology of using benchmarking indicators as performance indicators.

Another major line of future research relates to the implications of neo-liberalism on various market segments of the WSS sector, and the interrelations between the different markets. The segmentation of the WSS sector into four main markets opens a window of research opportunities. This research specifically aimed to understand the implications in one market (the market to fulfil the governmental mandate), but new research may also enter the other 3 markets, and more specifically will try to identify the inter-relations between the different markets. So research could be directed towards understanding the influence of neo-liberalism on the market for water resources, for production inputs, or for customer service. Ultimately most importantly is to gain an understanding of the net effect of neo-liberal changes for the entire sector. The study at hand brings in this respect only one part of the puzzle. Hopefully future research will add the other pieces to complete the puzzle.

Strategy and performance of water supply and sanitation providers.
Effects of two decades of neo-liberalism

Bibliography

Bibliography

Aalberts, R.F.T., Dijkgraaf, E., Varkevisser, M. and H.R.J. Vollebergh, 2002. Welvaart en de regulering van netwerksectoren. OCFEB in opdracht van het Ministerie van Economische Zaken en het Ministerie van Financien.

Aitman, D., 2001. Competition law constraints on access charges in the England and Wales water industry. Utilities Policy 10 (2001), pp. 129–136.

Alaerts, G.J.F.R., 1995. Institutional Arrangements in the Water and Sanitation Sector, European Country Studies. IHE Working Paper, Delft.

Aldrich, H.E. 1979. Organizations and environments. Englewood Cliffs, NJ: Prentice-Hall

Alegre, H., 2000. Performance Indicators for Water Supply Services. IWA Publishing. London

Aluseau and PriceWaterhouseCoopers, 2003. Benchmarking des activites dans la gestion des eaux au grand-duche de Luxembourg, February 2003

Andrews, R., Boyne, G.A. and R.M. Walker, 2006. Strategy Content and Organizational Performance: An Empirical Analysis. Public Administration Review, 66, 1, 52-63.

Andrews, K.R., 1987. The concept of corporate strategy (3rd edition). Homewood, IL: Irwin.

Ansoff, H.I. and J.M. Stewart, 1967. Strategies for a technology-based business. Harvard Business Review 45 (6), pp 71-83.

Armstrong, M., Cowan, S. and J. Vickers, 1994. Regulatory Reform: Economic Analysis and British Experience, MIT Press. Cambridge, MA.

Asociación Espanola De Abastecimientos De Agua y Saneamiento, 2003. Encuesta de Tarifas 2002-AEAS-AGA.

Asociación Espanola De Agua y Saneamiento, 2000. Suministro de agua potable y saneamiento en España. VII Encuesta Nacional de abastecimiento saneamiento y depuraciones.

Assies, W., 2003. David versus Goliath in Cochabamba: Water rights: neoliberalism, and the revival of social protest in Bolivia. Latin American Perspectives, 30(3), 14-36.

Bahaee, M.S., 1992. Strategy-comprehensiveness fit and performance. Australian Journal of Management. 17, 2, December 1992. The University of New South Wales.

Bain, J.S., 1968. Industrial Organization, John Wiley, New York.

Bakker, K., 2003b. Achipelagos and networks: urbanization and water privatization in the South. The Geographical Journal, 169(4), 328–341.

Banca Intesa, 2003, L'industria dei servizi idrici. Studi di settore.

Baptista, J.M., Passáro D.A. and R. Ferreira dos Santos, 2003. As linhas estratégicas do modelo de regulação a implementar pelo instituto regulador deáguas e residuos. IRAR.

Barraque, B., Grand d'Esnon, A. and P. van de Vyver, 2000. Experiences in France, in Holzwarth F., Kraemer R.A.

Becker, H.A. and J.W.M. van Doorn, 1987. Scenarios in an organisational perspective. Futures, December 1987. Buttersworth & Co (Publishers) Ltd.

Berg, S. and L. Holt, 2002. Strategic Considerations for Improving Water Sector Performance. Water21, August 2002, 51-54.

Berg, J. van den, and B.L.J.M. van de Reyt, 1996. The water sector. Economisch Bureau.

Bhattacharyya, A., Harris, T., Narayanan, R. and K. Raffiee, 1995. Specification And Estimation Of The Effect Of Ownership On The Economic Efficiency Of The Water Utilities. Regional Science and Urban Economics 25: 759-784.

Bhattacharyya, A., Parker, E. and K. Raffiee, 1994. An Examination of the Effect of Ownership on the Relative Efficiency of Public and Private Water Utilities, Land Economics 70.2: 197-209.

Blokland, M., Braadbaart, O., and K. Schwartz, 1999. Private Business, public owners. Government shareholdings in water enterprises. The Ministry of Housing, Spatial Planning, and the Environment. The Netherlands.

Boland, J.J., 1983. Water and wastewater pricing and financial practices in the U.S. metametrics. Washington, DC.

Boogaard, E. van den, and K. van Dijken, 2006. Trends bij gemeentelijke rioleringszorg worden zichtbaar. H2O 2, 2006. pp. 8-11.

Booker, A., 1994. British privatisation: balancing needs. Journal AWWA, March 1994.

Boubakri, N. and J. Cosset, 1998. The Financial and Operating Performance of Newly Privatised Firms: Evidence from developing countries. Journal of Finance, 53. 1081-1110

Boyer, M. and S. Garcia, 2002. Organisation et réglementation des services publics d'eau potable et d'assainissement en France. Cahiers de Recherche CIRANO. Rapport de projet, n° 2002 RP-13.

Boyne, G.A. and R.M. Walker, 2004. Strategy Content and Public Service Organizations. Journal of Public Administration Research and Theory 14, 2, pp. 231-252.

Boyne, G.A., 2003. Sources of Public Service Improvement: A Critical Review and Research Agenda. Journal of Public Administration Research and Theory 13: 367-394

Braadbaart, O., 2005. Privatizing water and wastewater in developing countries: assessing the 1990s'experiments. Water Policy No. 7

Bracker, J., 1980. The Historical Development of the Strategic Management Concept. Academy of Management Review, Vol. 5, No. 2 (Apr., 1980), pp. 219-224

Brdjanovic, D. and H. Gijzen, 2005. Challenges in Achieving a Sustainable Water Supply and Sanitation Services for Small Islands: the Caribbean Perspective. In proceedings Aqua 2005, Cali, Colombia.

Brdjanovic, D., Schouten, M., Hes, E, Schippers, J., Kennedy, M., Trifunovic, N., Vojinovic, Z., Van der Steen, P., Von Münch, E., Van Dijk, M.P., De Bruijn, R. and J. Gupta, 2005. Evaluation of the Proposals for an Integrated Sewage and Drinking Water System on St. Maarten. Vol.1

Bressers, H.A., Huitema, D. and S.M.M. Kuks, 1993. Policy Networks in Dutch Water Policy. Paper for the international conference CSTM 'Water policy networks in four countries', University of Twente, Enschede, The Netherlands October 1993

Brinkhorst, L.J., 2005. Letter to the Chairman of the Dutch Parliament addressing questions from member Van Bommel. 23 August 2005. E/EM/5050358

Briscoe, J., 1997. Managing water as an economic good: rules for reformers. Water Supply. Vol. 15, no. 4, pp. 153-172. 1997

Brown, A., 2002. Confusing means and ends, framework of restructuring, not privatization, matters most. International Journal of Regulation and Governance, 1(2), 115-128.

Brown, W.A. and J.O. Iverson, 2004. Exploring strategy and board structure in nonprofit organizations, Nonprofit and Voluntary Sector Quarterly, Sep, Vol. 33, pp.377–400.

Bruggink, T.H, 1982. Public versus Regulated Private Enterprise in the Municipal Water Industry: A Comparison of Operating Costs. Quarterly Review of Economics and Business 22 (Spring 1982): 11 1-25.

Buckland, J. and T. Zabel, 1998. Economic and financial aspects of water management policies In: Correia FN (ed.) Selected Issues in Water Resources Management in Europe, AA Balkema, 261-318

Budds, J. and G. McGranahan, 2003. Privatization and the provision of urban water and sanitation in Africa, Asia and Latin America. Human Settlements Discussion Paper Series. Theme: Water-1.

Bush, R.J. and S.A. Sinclair, 1991. A multivariate model and analysis of competitive strategy in the U.S. hardwood lumber industry. Forest Science. Vol. 37, No 2, pp. 481-499.

Byatt, I., 1993. Regulating a privatised water industry. Office of Water Services, UK

Byrnes, P., 1991. Estimation of Cost Frontiers In The Presence Of Selectivity Bias: Ownership and Efficiency of Water Utilities." In Advances in Econometrics edited by G. Rhodes. Vol. 9 (1991): 121-137.

Byrnes, P., Grosskopf, S. and K. Hayes, 1986. Efficiency and Ownership: Further Evidence. The Review of Economics and Statistics, Vol. 68, No. 2. (May, 1986), pp. 337-341.

Camacho, H., 2005. Water privatization policies and concession arrangements in Bolivia: Case study of La Paz and Cochabamba. MSc Thesis WM 05.08, Delft: UNESCO-IHE Institute for Water Education.

Camdessus, M. and J. Winpenney, 2003. Financing Water for All. Report of the World Panel on Financing Water Infrastructure.

Carney, M., 1990. Privatisation of water authorities in England & Wales. AWWA Annual Conference, Cincinnati. Public Officials Session.

Chandler, A.D., Jr, 1962. Strategy and Structure: Concepts in the History of the Industrial Enterprise. MIT Press, Casender, MA.

Child, J., 1972. Organizational Structure, Environment, and Performance-The Role of Strategic Choice. Sociology 6: 1-22.

Clarke, R.G. and S.J. Wallsten, 2002. 'Universal(ly Bad) Service: Providing Infrastructure Services to Rural and Poor Urban Consumers', Policy Research Paper 2868, Washington, DC.: World Bank.

Cmnd 9734, 1986. White paper on privatisation of the water authorities in England and Wales. HMSO. London.

Comitato Di Vigilanza Sull'Uso Delle Risorse Idriche, 2003, Relazione annuale al Parlamento sullo stato dei servizi idrici Anno 2002. www.minambiente.it

Conant, J.S., Mokwa, M.P, and P.R. Varadarajan, 1990. Strategic types, distinctive marketing competencies and organisational performance: a multiple measures-based study. Strategic Management Journal, Vol. 11, pp. 365-383.

COVIRI, 2004. Terzo Rapporto sullo stato di Avanzamento della Legge 5 Gennaio 1994. No. 36. Comitato per la vigilanza suli'uso delle risporse idriche. Roma, Luglio.

Cowan, S., 1997. Competition in the water industry. Oxford review of economic policy, vol.13 n.1, 83-92.

Coyle, G., 1997. The nature and value of futures studies or do futures have a future? Futures, Vol 29, No. 1, pp. 77-93. Elsevier Science Ltd.

Crain, M. and A. Zardkoohi, 1978. A Test of the Property Rights Theory of the Firm: Water Utilities in the United States. Journal of Law and Economics, October 1978.

Croteau, A. and F. Bergeron, 2001.An information technology trilogy: business strategy, technological deployment and organizational performance. The Journal of Strategic Information Systems. Volume 10, Issue 2, June 2001, Pages 77-99

Croteau, A., Raymond, L. and F. Bergeron, 1999. Testing the validity of Miles and Snow's typology. Academy of Information and Management Sciences Journal, Volume 2, Number 2.

Dalhuisen, J.M., 2003. The Netherlands. Aqualibrium project.

Dalhuisen, J.M., De Groot, H. and P. Nijkamp, 1999. The economics of water. Vrije Universiteit Amsterdam. Research memorandum 1999-36

Deakin, N. and K. Walsh, 1996. The enabling state, the role of markets and contracts. Public Administration, Vol. 74 Spring, 33-48.

Defra, 2003. The environment in your pocket 2003, London: Department for Environment, Food and Rural Affairs (Defra)

Defra, 2002. Sewage Treatment in the UK, UK Implementation of the EC Urban Waste Water Treatment Directive. London: Department for Environment, Food and Rural Affairs (Defra)

Delegation a l'Amenagement et au Developpement Durable du Territoire, 2003, Rapport d'information sur la gestion de l'eau sur le territoire, n. 1170: www.assemblee-nat.fr/12/pdf/rap-info/i1170.pdf

Demsetz, H., 1968. Why regulate utilities? Journal of Law and Economics, vol. 11, pp. 55-65.

Desarbo, W.E., Di Benedetto, C.A., Song, M. and I. Sinha, 2005. Revisiting the Miles and Snow strategic framework: uncovering relationships between strategic types, capabilities, environmental uncertainty, and firm performance. Strategic Management Journal, Volume 26: page 47-74. John Wiley & Sons, Ltd.

Dewenter, K. and P. Malatesta, 2001. State owned and privately owned firms: an empirical analysis of profitability, leverage and labour intensity. American Economic Review.

Dijk, M.P. van, and K. Schwartz, 2005. New public management en de Nederlandse drinkwatersector: nattigheid op globale, nationale en locale schaal. Economen Blad, 28 (Sept), 17-20

Dijk, M.P. van, 2003. Liberalisation of Drinking Water in Europe and Developing Countries. Inaugural Address, UNESCO-IHE Delft

Dillman, D.A., 2000. Mail and Internet Surveys – The Tailored Design Method. Wiley & Sons, New York

Donahue, J., 1989. The Privatization Decision: Public Ends, Private Means. New York: Basic Books.

Doty, D.H., Glick, W.H. and G.P. Huber, 1993. Fit, Equifinality, and Organizational Effectiveness: A Test of Two Configurational Theories. The Academy of Management Journal, Vol. 36, No. 6. (Dec., 1993), pp. 1196-1250.

DREE - Mission Economique de l'Ambassade de France, 2003, Le secteur de l'eau à usage domestique au Portugal. Fiche de sinthèse : www.dree.org/portugal

DWWA, 2001. Website information. Danish water and wastewater association (DWWA): http://danva.dk/sw114.asp

Eekeren, M. van, 2002. Toekomstverkenningen voor de waterleidingbedrijfstak. H2O, 21-2002

Estache, A. and M.A. Rossi, 1999. Comparing the Performance of Public and Private Water Companies in Asia and Pacific Region: What a Stochastic Costs Frontier Shows. Policy Research Working Paper, World Bank

Estache, A. and M.A. Rossi, 2002. How Different Is the Efficiency of Public and Private Water Companies in Asia? The World Bank Economic Review, Vol. 16, No. 1, 139-148. The World Bank

Estache, A. and E. Kouassi, 2002. Sector organization, governance, and the inefficiency of African water utilities. World Bank Policy Research Working Paper 2890.

EUREAU, 1997. Management Systems of Drinking water Production and Distribution Services in the EC Member States 1996

EUREAU, 1993. Management Systems of Drinking water Production and Distribution Services in the EC Member States 1992

EUROMARKET, 2005. Workpackage Deliverable 4 Comparative Policy Analysis of Legislation in the Water and Sanitation Sector in Europe. Euromarket

European Commission, 1997. Amendment of the Environmental Protection Act of 1991, section 3

Eurostat, 2007. Quarterly Panorama of European Business Statistics 3-2007. European Commission.

Faria, R.H., Da Silva Souza, G. and T.B. Moreira, 2005. Public versus private water utilities: empirical evidence for Brazilian companies. Economics Bulletin, Vol. 9. No. 2, pp. 1-7.

Feigenbaum, S. and R. Teeples, 1983. Public versus Private Water Delivery: A Hedonic Cost Approach. Review of Economics and Statistics 65 (November 1983): 672-78.

Finger, M., Allouche, J. and P. Luís-Manso, 2007. Water and Liberalisation: European Water Scenarios. European Commission Community Research. IWA Publishing.

Fink, A., 1995. The survey kit. Thousand Oaks: SAGE Publications, Inc.

Finnegan, W., 2002. Leasing the Rain. The New Yorker.

Föllmi, R. and U. Meister, 2002. Product Market Competition in the Water Industry: Voluntary Nondiscriminatory pricing. Institute for Empirical Research in Economics. University of Zurich. Working Paper Series. Working Paper No. 115. May 2002

Fonseca, C., Schouten, M. and R. Franceys, 2005. Plugging the Leak. Can Europeans find new sources of funding to fill the MDG water and sanitation gap? IRC. June 2005. 54 pages.

Foster, V., 1996. Policy Issues for the Water and Sanitation Sectors. Washington D.C.: IDB.

Fox, W., and R. Hofler, 1986. Using Homothetic Composed Error Frontiers to Measure Water Utility Efficiency. Southern Economic Journal 53(2): 461-77.

Fox-Wolfgram, S.J., Boal, K.B. and J.G. Hunt, 1998. Organizational adaptation to institutional change: A comparative study of first-order change in prospector and defender banks. Administrative Science Quarterly, 43: pp. 87-126.

Francisco, N., 1998. Institutions for Water resources Management in Europe.

Freeman, R.E. and P. Lorange, 1985. Theory Building in Strategic Management, in R. Lamb and P. Shrivastava (eds.), Advances in Strategic Management, Volume 3, 1985, Greenwich: JAI Press, 9-38.

Freeman, C., 1974. The economics of industrial innovation. Penguin. Harmondsworth. England.

Fudenberg, D. and J. Tirole, 1991. Game theory. Massachusetts Institute of Technology

Galal, A., Jones, L., Tandon, P. and I. Vogelsang, 1994. Welfare consequences of selling public enterprises. World Bank, Washington, D.C.

Gimenez, F.A.P., 2000. The Benefits of a Coherent Strategy for Innovation and Corporate Change: A Study Applying Miles and Snow's Model in the Context of Small Firms. Creativity and Innovation Management 9 (4), 235–244.

Glaister, S., 1996. Incentives in Natural Monopoly: the case of water. In M.E. Beesley (ed.) Regulating utilities: a time for change?, IEA Readings 44, May 1996

Gleick, P.H., Wolff, G., Chalecki, E.L. and R. Reyes, 2002. The New Economy of Water: The Risks and Benefits of Globalization and Privatization of Fresh Water, Oakland: Pacific Institute for Studies in Development, Environment, and Security.

Grabowski & Poort, St. Maarten N.V., 1998. Ontwerp Eilandelijk Rioleringsplan St. Maarten 1998-2010

Grabowski & Poort St. Maarten N.V., 1995. Basis Riolerings Plan St. Maarten

Green, C., 2000. If only life were that simple; optimism and pessimism in Economics. Physics and Chemistry of the Earth. Vol 25, No. 3, pp. 20-21

Haarmeyer, D. and A. Mody, 1998. Financing water and sanitation projects, the unique risk. Public Policy for the Private Sector, Note no. 151. Washington D.C.: World Bank.

Hall, D., 2007. Self-assessment or public debate? -evaluating the liberalisation of network services in the EU and USA. A report commissioned by the European Federation of Public Service Unions (EPSU). http://www.psiru.org/reports/2007-09-EU-HEPNI.doc

Hall, D., 2003. Water and DG Competition. PSIRU.

Hambrick, D.C. and S. Finkelstein, 1987. Managerial discretion: A bridge between polar views of organizational outcomes. Research in Organizational Behavior, 9, pp. 369-406.

Hambrick, D.C., 1983. High profit strategies in mature capital goods industries: a contingency approach. Academy of Management Journal, 26. pp. 687-707

Hambrick, D.C., 1980. Operationalizing the concept of business-level strategy in research. Academy of Management Review, Vol. 5. No. 4, 567- 575

Hannan, M. and J. Freeman, 1977. The population ecology of organizations. American Journal of Sociology, 82: 929-964.

Hansen, W., Herbke, N., Kraemer, R.A., Oppolzer, G. and W. Schönbäck, 2003. Internationaler Vergleich der Siedlungswasserwirtschaft. Band Länderstudien Österreich, England und Wales, Frankreich Überblicksdarstellungen Deutschland und Niederlande. Ecologic and Ifip TU, Berlin-Brüssel-Wien.

Hansen, G. and B. Wernerfeldt, 1989. Determinants of firm performance: the relative importance of economic and organisational factors. Strategic Management Journal 10: 399-411

Hart, O., 2003. Incomplete contracts and public ownership: remarks and an application to public–private partnerships, Economic Journal 119 (2003), pp. C69–C76.

Hart, S., 1992. An Integrative Framework for Strategy-Making Processes. Academy of Management Review, Vol. 17, No. 2, pp. 327-351.

Hart-Landsberg, M., 2006. Neoliberalism: Myths and Reality. Monthly Review. New York: Apr 2006. Vol. 57, Iss. 11; pg. 1-17

Harvey, D., 2005. A brief history of neoliberalism. Oxford University Press Inc., New York.

Hatch, M.J., 1997. Organizational Theory: Modern, Symbolic and Postmodern Perspectives. Oxford: Oxford University Press Inc., New York.

Haut Conseil du Secteur Public, 1999. Quelle regulation pour l'eau et les services urbaines? La documentation française.

Hilderbrand, M., 2002. Water for El Alto: Expanding Water and Sanitation for the Poor. Cambridge: Kennedy School of Government.

Holzwarth, F. and R.A. Kraemer, 2000, Umweltaspekte einer Privatisierung der Wasserwirtschaft in Deutschland. Ecologic.

IFEN, 2003. Le comptes economiques de l'environment en 2001.

INAG, 2002. Plano National da Agua. INAG: www.inag.pt

Israel, A., 1994. Dealing with gaps in public sector performance: institutional development experience of the World Bank. In: Uphoff, N. (ed.), Puzzles of Productivity in Public Organisations, San Francisco: ICS Press.

IWA, 2003. International Water Utility Leaders Forum, charters on sustainable water management, governance and regulation. 1st draft.

Jauch, L.R., and R.N. Osborn, 1981. Toward an Integrated Theory of Strategy. The Academy of Management Review, Vol. 6, No. 3. (Jul., 1981), pp. 491-498.

Jeffery, J., 1994. Privatization in England and Wales. Journal AWWA. March 1994.

Jeffery, M., 1990. Privatization in the UK water industry. The Thames Experience. Thames Water International.

Jerome, A., 2004. Privatisation and Regulation in South Africa. An evaluation. 3rd International Conference on Pro-Poor regulation and competition: issues, policies and practices. Cape Town – South Africa. 7-9 September 2004.

Johnson, J. and K. Scholes, 1988. Exploring Corporate Strategy, 2nd ed. Prentice Hall, New York.

Joyce, P., 1992. Strategic Management for the Public Services. Open University Press, Buckingham, USA

Kallis, G. and P. Nijkamp, 1999. Evolution of European Union water policy: a critical assessment and a hopeful perspective, Free University Research memorandum, n° 27, Amsterdam: Vrije Universiteit.

Kaplan, R. and D. Norton, 2004. Strategy Maps: Converting Intangible Assets into Tangible Outcomes. Harvard Business School Press, USA.

Karimi, J., Gupta, Y.P. and T.M. Somers, 1996. Impact of competitive strategy and information technology maturity on firm's strategic response to globalization. Journal of Management Information Systems 12 (4), pp. 55-88

Katko T., 2000. Long-term development of water and sewage services in Finland. Public Works Management & Policy. APWA . Sage Publications. Vol. 4, no. 4. pp. 305-318.

Kessides, I., 2003. Reforming infrastructure: privatization, regulation and competition. World Bank Policy Research Report. Washington D.C.: World Bank.

Ketchen, D.J., Combs, J.G., Russell, C.J., Shook, C., Dean, M.A., and J. Runge, 1997. Organizational configurations and performance: A meta-analysis. Academy of Management Journal, 40, 223-240.

Kirkpatrick, C., Parker, D. and Y. Zhang, 2004. State versus private sector provision of water services in Africa: A statistical, DEA and stochastic cost frontier analysis. Centre on Regulation and Competition. Working Paper Series, 70. Institute for Development Policy and Management. University of Manchester.

Klostermann, J.E.M., 2003. The Social Construction of Sustainability in Dutch Water Companies. Dissertation. Erasmus Universiteit Rotterdam, December.

Komives, K., 1999. Designing pro-poor water and sewer concessions: Early lessons from the Aguas del Illimani concession in Bolivia. Washington, DC: World Bank.

Kotler, P., 1998. Marketing Management: Analysis, Planning, Implementation and Control. Prentice-Hall. New Jersey, USA.

Kuks, S.M.M., 2006. The privatisation debate on water services in the Netherlands: public performance of the water sector and the implications of market forces, Water Policy 8, pp. 147-169

Kuks, S.M.M., 2003. The privatisation debate on water services in the Netherlands: an examination of the public duty of the Dutch water sector and the implications of market forces and water chain cooperation. University of Twente, Enschede, The Netherlands.

Lambert, D., Dichey, D. and K. Raffiee, 1993. Ownership and Sources of Inefficiency in the Provision of Water Services, Water Resources Research 29.6: 1,573-78.

Lannerstad, M., 2003, AQUALIBRIUM - European Water Management between Regulation and Competition - The Swedish Country Report, Published in: Mohajeri, S., Knothe, B., Lamothe, D-N., Faby J-A, Aqualibrium – European Water Management between Regulation and Competition, Directorate-General for Research: Global Change and Ecosystems, European Commission, Brussels, p. 261-287, 2003

Lipsey, R.G., Steiner, P.O. and D.D. Purvis, 1987. Economics. Harper & Row Publishers. New York. USA.

Llewellyn, S. and E. Tappin, 2003. Strategy in the public sector: management in the wilderness. Journal of Management Studies 40:4 June 2003. Blackwell Publishing, USA.

Lobina, E., 2005. D10f: WaterTime National Context Report – Italy, WaterTime Deliverable D10f (http://www.watertime.net/docs/WP1/NCR/D10f_Italy.doc).

Lynk, E.L., 1993. Privatisation, Joint Production and the Comparative Efficiencies of Private and Public Ownership: The UK Water Industry Case. Fiscal Studies 14 (2), 98–116.

Mann, P.C. and J.L. Mikesell, 1976. Ownership and Water System Operation. Journal of the American Water Resources Association 12 (5), 995–1004

Martin, S. and D. Parker, 1997. The Impact of Privatization—Ownership and Corporate Performance in the UK, Routledge, London.

Martins, R. and A. Fortunato, 2002. Regulatory Framework in Water and Sewerage Services in Portugal. Paper presented at the 6th EUNIP Annual Conference, Abo/Turku, 5th – 7th December 2002.

MAOT, 2000, Plano Estratégico de Abastecimento de Água e de Saneamento de Águas Residuais (2000-2006). Lisboa, Ministério do Ambiente e do Ordenamento do Território.

Marra, A., 2006. Mixed Public-Private Enterprises in Europe: Economic Theory and an Empirical Analysis of Italian Water Utilities. Bruges European Economic Research Papers. BEER paper n° 4. July 2006. http://www.coleurop.be/eco/publications.htm

Mason, E.S., 1957. Economic Concentration and the Monopoly Problem, Harvard University Press, Cambridge, MA.

Massarutto, A., 2000, Experiences in Italy, in Holzwarth F., Kraemer R.A, 2000.

Matthews, R.C.O., 1986. The Economics of Institutions and the Sources of Growth. Economic Journal, 96: 903-918.

McCarthy, J. and S. Prudham, 2004. Neoliberal nature and the nature of neoliberalism. Geoforum. Volume 35, Issue 3, May 2004, Pages 275-283

McCloskey, D.N., 1998. The Rhetoric of Economics. 2d ed. Madison, Wis.: University of Wisconsin Press

McDaniel, S.W. and J.W. Kolari, 1987, Marketing Strategy Implications of the Miles and Snow Strategic Typology. Journal of Marketing, 51 (4), 19-30.

Megginson, W.L. and J.M. Netter, 2001. From State to Market: A survey of empirical studies on privatisation. Journal of Economic Literature. Vol. 39, June. pp. 321-89

Melosi, M.V., 2004. Full Circle: Public Goods versus Privatization of Water Supplies in the United States. International Summer Academy for Technology Studies – Urban Infrastructure in Transition.

Memon, F.A. and D. Butler, 2003a. The role of privatisation in the water sector. Water Perspectives. No. 1. March. Page 28-36

Memon, F.A. and D. Butler, 2003b. European Water Management between Regulation and Competition. Chapter United Kingdom. Aqualibrium. European Commission Community Research.

Ménard, C. and S. Saussier, 2000. Contractual Choice and Performance: The Case of Water Supply in France. Revue D'économie Industrielle. 92(2-3) (2000): 385-404.

Meyer, A.D., 1982. Adapting to environmental jolts. Administrative Science Quarterly 27, pp. 515-537.

Miles, R.E. and C.C. Snow, 1978. Organisational strategy, structure and process. New York, McGraw-Hill

Miller, D., 1988. Relating Porter's Business Strategies to Environment and Structure: Analysis and Performance Implications. The Academy of Management Journal, Vol. 31, No. 2 (Jun., 1988), pp. 280-308. Published by: Academy of Management

Miller, D. and P.H. Friesen, 1977. Strategy-making in context: ten empirical archetypes. Journal of Management Studies, 14 (3) 253-280.

Mintzberg, H., Ahstrand, B. and J. Lampel, 1998. Strategy safari: a guided tour through the wilds of strategy. New York: The Free Press.

Mintzberg, H., 1987a. The strategy concept I: Five P's for strategy. California Management Review, 30: pp 11 -24.

Mintzberg, H., 1987b. Crafting Strategy, Harvard Business Review, July-August, 66-75.

Mintzberg, H., 1973. The Nature of Managerial Work, Harper and Roe, New York.

Morgan, W.D., 1977. Investor Owned vs. Publicly Owned Water Agencies: An Evaluation of the Property Rights Theory of the Firm. Water Resources Bulletin 13 (August 1977): 775-81.

Naismith, I., 2003. Taking a big step into sewage. Water and Waste Treatment. Vol. 46, no. 6, pp. 22-24. June 2003

Namiki, N., 1989. Miles and Snow typology of strategy, perceived environmental uncertainty, and organizational performance. Akron Business and Economic Review 20 (2), pp. 72-88.

Neal, K., Maloney, P.J., Marson, J.A. and T.E. Francis, 1996. Restructuring America's Water Industry: Comparing Investor-owned and Government Water Systems. Reason Public Policy Institute. Policy Study No. 200. January.

Nickson, A. and D. Muscoe, 2004. Analysis of the European Water Supply and Sanitation Markets and its possible evolution. Annex 3. Final report Workpackage 2 Euromarket.

Nickson, A. and C. Vargas, 2002. The limitations of water regulation: The failure of the Cochabamba concession in Bolivia. Bulletin of Latin American Research, 21(1), 128-149.

Nickson, A., 1997. The public-private mix in urban water supply, International Review of Administrative Sciences, vol. 63, 165-186.

Njiru, C. and K. Sansom, 2003. Strategic marketing of water services in developing countries. Proceedings of the Institution of Civil Engineers, Municipal Engineer, 156, No. 2, 143-148.

Noll, R., Shirley, M. and S. Cowan, 2000. Reforming urban water systems in developing countries, SIEPR Discussion Paper No. 99-32. Stanford: Stanford Institute for Economic Policy Research.

North, D.C., 1990. Institutions, Institutional Change and Economic Performance. Political Economy of Institutions and Decisions. Cambridge University Press.

NRC, 1995. Measuring and Improving Infrastructure Performance, Committee on Measuring and Improving Infrastructure Performance, Board on Infrastructure and the Constructed Environment, National Research Commission (NRC). Washington, DC, National Academy Press.

OECD, 2003. Water partnerships: Striking a balance. 7 April, OECD Observer, Paris: OECD

OECD, 1998. Water, Performance and Challenges in OECD countries. OECD

OFWAT, 2007. International Comparison of Water and Sewerage Service: 2004–05, Office of Water Services, Birmingham (April 2007).

OFWAT, 2006. International Comparison of Water and Sewerage Service: 2003-04 Report, Office of Water Services, Birmingham.

OFWAT, 2005. International Comparison of Water and Sewerage Service: 2002–03 Report, Office of Water Services, Birmingham (March 2005).

OFWAT, 2004. International Comparison of Water and Sewerage Service: 2001–02, Office of Water Services, Birmingham (March 2004).

OFWAT, 2002a. International Comparison of Water and Sewerage Service: 2000–01 Report, Office of Water Services, Birmingham (December 2002).

OFWAT, 2002b. Tariff Structure and Charges: 2002 – 2003 Report, published May 2002, Birmingham: OFWAT. Office of Water Services.

OFWAT, 2001. World Wide Water Comparisons: 1999-2000 Report. Office of Water Services, Birmingham.

OFWAT, 2000. Comparing the performance of the water companies in England and Wales in 1998-99 with water enterprises in other industrialised countries. Office of Water Services, Birmingham.

OFWAT, 1999. Comparing the performance of the water companies in England and Wales in 1998-99 with water enterprises in other industrialised countries. Office of Water Services, Birmingham.

Ogden, S.G., 1995. Transforming frameworks of accountability: the case of water privatization. Accounting Organisations and Society. Vol. 20. No. 2. Elsevier Science. Great Britain.

Okten, C. and K.P. Arin, 2006. The Effects of privatization on Efficiency: How Does Privatization Work? World Development, Vol. 34(9), pp. 1537-1556.

Oppenheim, A.N., 1992. Questionnaire design, interviewing and attitude measurement. New Edition. Cassell, London.

Pargal, S., 2003. Regulation and Private Sector Investment in Infrastructure: Evidence from Latin America. Policy Research Working Paper 3037. Washington DC: World Bank.

Parnell, A. and P. Wright, 1993. Generic Strategy and Performance: an Empirical Test of the Miles and Snow Typology. British Journal of Management 4 (1), 29–36.

Peck, J. and A. Tickell, 2002. Neoliberalizing Space. Antipode 34 (3), 380–404.

Perdok, P.J. and J. Wessel, 1998. The Netherlands. In: Institutions fir water resources management in Europe, edited by Nunes Correia. Volume 1. A.A. Balkema.

Pielen, B., Herbke, N., and E. Interwies, 2004. Analysis of the legislation and emerging regulation at the EU country level. Country Report United Kingdom. Chapter 9 of the Deliverable 4 of the Euromarket project.

Pinsent Masons, 2008. Pinsent Masons Water Yearbook 2008-2009. Published by Pinsent Masons, London.

Poister, T., and G. Streib, 1999. Performance measurement in municipal government: Assessing the state of the practice. Public Administration Review, 59, 325-335.

Pollit, C. and G. Bouckaert, 2000. Public Management Reform, a Comparative Analysis. Oxford University Press. Oxford

Porter, M.E., 1996. What is Strategy? Harvard Business Review, November - December.

Porter, M.E., 1981. The contributions of industrial organization to strategic management. Academy of Management Review, 6, 1981, pp. 609-620.

Porter, M.E., 1980. Competitive Strategy. Free Press, New York

Prescott, J.E., 1986. Environments as moderators of the relationship between strategy and performance. Academy of Management Journal, 29(2), pp. 329-346

Publius Ovidius Naso, 43 B.C. *Ars Amatoria.*

Raffiee, K., Narayanan, R., Harris, T., Lambert, D. and J. Collins, 1993. Cost Analysis of Water Utilities: A Goodness-of-Fit Approach, Atlantic Economic Journal 21.3: 18-29.

Ramaswamy, K., Thomas, A.S. and R.J. Litschert, 1994. Organizational Performance in a Regulated Environment: The Role of Strategic Orientation. Strategic Management Journal, Vol. 15, No. 1. (Jan., 1994), pp. 63-74.

Reger, R.K., Duhaime, I.M. and J.L. Stimpert, 1992. Deregulation, Strategic Choice, Risk and Financial Performance. Strategic Management Journal 13: 189-204

Renzetti, S. and D. Dupont, 2004. The performance of municipal water utilities: evidence on the role of ownership. Journal of Toxicology and Environmental Health, Part A, Volume 67, Issue 20 - 22 January 2004 , pages 1861 - 1878

Ring, P.S. and J.L. Perry, 1985. Strategic management in public and private organizations: implications and distinctive contexts and constraints, Academy of Management Review 10, 276-286

RIONED, 2002. Riool in cijfers. http://www.Rioned.org.

Robinson, C., 1997. Introducing competition into water. In: Beesley, M.E. (ed.) Regulating utilities: broadening the debate. IEA Readings.

Rond, M. de, and R.A. Thietart, 2007. Choice, chance, and inevitability in strategy. Strategic Management Journal, 28(5): 535-551.

Rossmann, B., 2001. Finanzierung der Investitionen in der Siedlungswasserwirtschaft. In Österreichischer Städte- und Gemeindebund: Finanzausgleich 2001. Das Handbuch für die Praxis, Wien

Rudolph, K.U. (ed.), 2001. Policies and Experiences. Federal Ministry for the Environment, Nature Conservation and Nuclear Society, Federal Environmental Agency.

Ruekert, R.W. and O. Walker, 1987. Marketing's Interaction with Other Functional Units: A Conceptual Framework and Empirical Evidence. Journal of Marketing 51 (1), pp. 1–19.

Saal, D. and D. Parker, 2001. Productivity and Price Performance in the Privatized Water and Sewerage Companies of England and Wales. Journal of Regulatory Economics. 20(1) (2001): 61-90.

Sachs, J., Zinnes, C. and Y. Eilat, 2000. The gains from Privatization in Transition economies: is 'change of ownership' enough? Consulting Assistance on Economic Reform II Discussion Paper 63. Harvard University, Center for International Development, Cambridge, Mass.

Savenije, H., 2002. Why water is not an ordinary economic good, or why the girl is special. Physics and Chemistry of the Earth. pp. 741 - 744

Sawkins, J., 2001. The Development of Competition in the English and Welsh Water and Sewerage Industry. Fiscal Studies, 22, 2, 189-215.

Scharfenaker, M.A., 1992. The politics of water supply, Journal AWWA, August.

Scherer, F.M., 1980. Industrial Market Structure and Economic Performance. Houghton Mifflin Company, Boston, MA.

Schouten, M. and M.P. van Dijk, forthcoming (accepted May 2008). Regulation and comparative discretion of publicly and privately owned water companies in the Netherlands, England and Wales. Water Policy. IWA Publishing.

Schouten, M. and M.P. van Dijk, 2008. Private Sector Involvement according to European Water Liberalisation Scenarios. International Journal of Water. Vol. 4, No. 3/4: 180-196. Inderscience Enterprises Ltd.

Schouten, M., Brdjanovic, D. and M.P. van Dijk, 2008. A Caribbean Evaluation of Public versus Private Drinking Water Provision: the case of St. Maarten, Netherlands Antilles. International Journal for Water. Vol. 4, No. 3/4: 258-274. Inderscience Enterprises. Ltd.

Schouten, M. and M.P. van Dijk, 2007. Chapter 1: The European Water Supply and Sanitation Markets. In: Finger, M., Allouche, J. and P. Luís-Manso (ed.). Water and Liberalisation: European Water Scenarios. European Commission Community Research. IWA Publishing.

Schouten, M., 2007. Exploring Strategies of Water Providers. Paper presentation at the 1st IWA Utilities Conference 'Customer Connection'. Maastricht, the Netherlands: 14-16 June 2007.

Schouten, M. and G. Casale, 2006. Strategie dei gestori idrico in Italia. Management delle Utilities. Anno 4-Numero 2. Rivesti Trimestrale. April-June 2006. Umbria.

Schouten, M. and K. Schwartz, 2006. Water as a political good: implications for investments. International Environmental Agreements: Politics, Law and Economics. Vol. 6. No. 2; pp. 407-421. Springer

Schouten, M. and M.P. van Dijk, 2005. Regulatory impositions posed upon strategic actions of publicly and privately owned water companies, in respectively the Netherlands and England and Wales. Contribution to International Symposium - 29-30 September 2005, University of Paris VIII – Saint Denis: "Competition and stakes in the regulation of services of general interest. Feedback of the last twenty years"

Schouten, M. and M.P. van Dijk, 2004a. The European Waterscape of Management Structures and Liberalisation. Water and Wastewater International Magazine. June 2004.

Schouten, M. and M.P. van Dijk, 2004b. The dynamics of the European Water Supply and Sanitation Market. AWRA International Specialty Conference 'Governance for People and Nature'. IWLRI. University of Dundee.

Schouten, M., Dijk, M.P. van, Swami, K. and M. Kooij, 2004. Analysis of the legislation and emerging regulation at the EU country level. Country Report United Kingdom. Chapter 4 of the Deliverable 4 of the Euromarket project.

Schwartz, K., 2006. Managing Public Water Utilities. An assessment of bureaucratic and New Public Management models in the water supply and sanitation sectors in low- and middle-income countries. PhD Thesis. UNESCO-IHE Delft, The Netherlands.

Schwartz, K.H. and M. Schouten, 2007. Water as a Political Good: Revisiting the Relationship between Politics and Service Provision. Water Policy. Vol. 9: No. 2; pp. 119-129. IWA Publishing.

Scottish Executive Environment Group, 2003. Draft Water Services Scotland Bill. Edinburgh.

Scottish Water, 2003. Annual Report 2002-2003. Dunfermline.

Segev, E., 1987. Strategy, Strategy making, and performance – an empirical investigation. Management Science. Vol. 33, No. 2, February 1987. USA. pp. 258-269

Seppala, C.T., Harris, T.J. and D.W. Bacon, 2001. Time Series Methods for Dynamic Analysis of Multiple Controlled Variables. Journal of Process Control. Volume 12, Issue 2, February 2002, Pages 257-276

Shaoul, J., 1997. A Critical Financial Analysis of the Performance of Privatized industries: The Case of the Water Industry in England and Wales." Critical Perspectives on Accounting. 8 (1997): 479-505.

Shirley, M. and P. Walsh, 2001. Public versus private ownership: the current state of the debate. Research paper World Bank, Washington

Shortell, S.M. and E.J. Zajac, 1990. Perceptual and archival measures of Miles and Snow's strategic types: a comprehensive assessment of reliability and validity. Academy of Management Journal, 33, 817–832.

Singh, J.V. and C.J. Lumsden, 1990. Theory and research in organizational ecology. American Review of Sociology, 16: 161-195.

Sisson, J.D., 1992. How strategic is your planning? Management and Organisations, Journal AWWA, November 1992.

Slater, S.F., Olson, E.M., Tomas, G. and M. Hult, 2006. The moderating influence of strategic orientation on the strategy formation capability-performance relationship. Strategic Management Journal. Volume 27, Issue 12 , Pages 1221 - 1231

Slater, S. and E. Olson, 2001. Marketing's contribution to the implementation of business strategy: an empirical analysis. Strategic Management Journal 22(11): pp. 1055–1068.

Smith, L., 2004. The murky waters of the second wave of neoliberalism: Corporatization as a service delivery model in Cape Town. Geoforum 35(3, May): 375-93.

Smith, K.G., Guthrie, J.P. and M. Chen, 1986. Miles and Snow's typology of strategy, organizational size and organizational performance. Academy of Management Proceedings, 1986, pp. 45-49

Snow, C.C. and D.C. Hambrick, 1980. Measuring organisational strategies: some theoretical and methodological problems. Academy of Management Review. Volume 5, no. 4. pp. 527-538

Snow, C.C. and L.G. Hrebiniak, 1980. Strategy distinctive competence and organizational performance. Administrative Science Quarterly 25, pp. 317-336

Speckbacher, G., 2003. The Economics of Performance Management in Nonprofit Organizations. Nonprofit Management and Leadership, vol. 13 n° 3, Spring 2003

Spiller, P. and W. Savedoff, 1999. Government opportunism and the provision of water. In: Spiller, P. and W. Savedoff (eds.), Spilled water, institutional commitment in the provision of water services. Washington D.C.: Inter-American Development Bank.

Stern, W. and G. Stalk, 1998. Perspectives on Strategy from the Boston Consulting Group N.Y.: John Wiley & Sons, Inc.

Stockholm Environmental Institute, 2003. Ministerial Conference on Environment and Sustainable Development, Lulea 28-29 August 2003, "Round table Discussion paper, Water supply and sanitation in the Nordic and Baltic and Barents regions"

Tavakolian, H., 1989. Linking the information technology structure with organizational competitive strategy. MIS Quarterly 13, pp. 309-317.

Teeples, R., Feigenbaum, S. and D. Glyer, 1986. Public Versus Private Water Delivery: Cost Comparisons. Claremont Center for Economic Policy Studies

Teeples, R. and D. Glyer, 1987. Cost of Water Delivery Systems: Specification and Ownership Effects. The Review of Economics and Statistics, Vol. 69, No. 3. (Aug., 1987), pp. 399-408.

Teulings, C.N., Bovenberg, A.L. and H.P. van Dalen, 2003. De calculus van het publieke belang. Kenniscentrum van Ordeningsvraagstukken. Publicatie Nr 03 ME 18. Den Haag.

Teulings, C., Van der Veen, R. and W. Trommel, 1997. Dilemma's van sociale zekerheid: een analyse van 10 jaar herziening van het stelsel van sociale zekerheid. 's Gravenhage, VUGA Uitgeverij, 220.

Thobani, M., 1999, Private Infrastructure, Public Risk, Finance and Development, March, World Bank.

Thomas, D.A. and R.R. Ford, 2005.The Crisis of Innovation in Water and Wastewater. Edward Elgar. Cheltenham, UK.

Thomas, A.S., Litschert, R.J. and K. Ramaswamy, 1991. The performance impact of strategy-manager coalignment: an empirical examination. Strategic Management Journal, Vol. 12, 509-522.

Truch, E. and D. Bridger, 2002. The importance of strategic fit in knowledge management, Proceedings of the 10th European Conference on Information Systems: "Information systems and the future of the digital economy" Editor: Wrycza, S.

Tsagarakis, K.P., Mara, D.D., Horan, N.J. and A.N. Angelakis, 2001. Institutional status and structure of wastewater quality management in Greece. Water Policy 3, 81-99.

Umwelt Bundes Amt for Humanity and Environment, 2002. Environmental data Germany.

UN, 2007. The Millennium Development Goals Report 2007. United Nations. New York

UNEP, 1999. Assessment of Land-based Sources and Activities Affecting Marine, Coastal and Associated Freshwater Environment in the Wider Caribbean Region, UNEP Regional Seas Reports and Studies No. 172.

Unie van Waterschappen, 1999. Bedrijfsvergelijking zuiveringsbeheer, Zuiver afvalwater.

Varone, F. and D. Aubin, 2002. The evaluation of the National water regime in Belgium. Euwareness.

Versteegh, J.F.M. and J.D. te Biesebeek, 2004. De kwaliteit van het drinkwater in Nederland, in 2002. VROM 3272. RIVM rapport 703719 005; www.rivm.nl

VEWIN, 2007. Water in zicht 2006. Bedrijfsvergelijking in de drinkwatersector. Vereniging van Waterbedrijven in Nederland (VEWIN). Rijswijk

VEWIN, 2004. Water in zicht 2003. Bedrijfsvergelijking in de drinkwatersector. Vereniging van Waterbedrijven in Nederland (VEWIN). Rijswijk.

VEWIN, 2003. Waterleidingstatistiek 2003. Netherlands Association water companies

VEWIN, 2001. Water supply statistics. http://www.vng.nl

V&W (Ministerie van Verkeer en Waterstaat), 1998. Vierde Nota Waterhuishouding.

Vickers, J. and G. Yarrow, 1991. Economic Perspectives on Privatisation. Journal of Economic Perspectives, 5 (2)

Vickers, J., 1995. Concepts of Competition. Oxford Economic Papers, New Series, Vol. 47, No. 1 (Jan., 1995), pp. 1-23. Published by: Oxford University Press

Villalonga, B., 2000. Privatization and efficiency: Differentiating ownership effects from political, organizational, and dynamic effects. Journal of Economic Behavior and Organization, 42: 43-74.

VnG, UvW and VEWIN, 2007. Samenwerken aan Water. Special van de Vereniging van Nederlandse Gemeenten (VnG), de Unie van Waterschappen (UvW) and de Verening van Water bedrijven (VEWIN). November 2006.

Voigt, S. and H. Engerer, 2002. Institutions and Transformation – Possible Policy Implications of the New Institutional Economics. In: Klaus F. Zimmermann (Ed.) Frontiers in Economics. Berlin: Springer Publishing Company, 127-184.

Wallsten, S., 2001. An econometric analysis of telecom competition, privatization and regulation in Africa and Latin America. Journal of Industrial Economics, 49 (1), pp. 1-20

Walsh, K., 1995. Public Services and Market Mechanisms. London, MacMillan

Ward, P.T., Duray, R., Leong, G.K and Chee-Chuong Sum, 1995. Business Environment, Operations Strategy, and Performance. An Empirical Study of Singapore Manufacturers. Journal of Operations Management, 13 (2), 1995. pp. 99-115.

Water Industry Commissioner for Scotland, 2001. Customer Service Report 2000-2001.

Wetenschappelijk Raad voor het Regeringsbeleid (WRR), 2000. Het borgen van publiek belang, Den Haag: SDU Uitgevers

WFD, 2000. Directive 2000/60/EC of the European Parliament and of the Council of 23 October 2000 establishing a framework for Community action in the field of water policy. Official Journal of the European Communities

White, R.E. and R.G. Hamermesh, 1981. Toward a model of business unit performance: An integrative approach. Academy of Management Review, 6, 213-223.

Williamson, O.E., 1998. The Institutions of Governance. The American Economic Review, Vol. 88, No. 2, Papers and Proceedings of the Hundred and Tenth Annual Meeting of the American Economic Association (May, 1998), pp. 75-79. Published by: American Economic Association

Williamson, J., 1990, What Washington Means by Policy Reform. In: Latin American Adjustment: How Much Has Happened? Edited by John Williamson (Washington: Institute for International Economics)

Wills-Johnson, N., Lowdon, A., and H. Bell, 2003. Access and competition in the water industry. Water Policy 5 (2003). IWA Publishing, London

Wit, B., de, and R. Meyer, 1994. Strategy Synthesis. Resolving Strategy Paradoxes to Create Competitive Advantage. International Thomson Business Press. UK.

Wood, R., 1999. The Future of Strategy: The Role of the New Science. In: Managing Complexity in Organisations: A View in Many Directions, edited by Lissack, R. & Gunz, H. P. London: Quorum Books. Page 5, 118 –164.

World Bank, 2006. Benchmarking Water & Sanitation Utilities: A Start Up Kit, transportation, water and urban development department: water and sanitation division World Bank

World Bank, 2005a. A Time to Choose. Caribbean Development in the 21st Century. April 26, 2005. Caribbean Country Management Unit, Poverty Reduction and Economic Management Unit. Latin America and the Caribbean Region

World Bank, 2005b. Achieving Environmentally Sustainable Tourism in the OECS Sub-Region. Washington, World Bank. Report No. 31725-LAC

World Bank, 2004. Public and Private Sector Roles in Water Supply and Sanitation Services. Operational Guidance for World Bank Group Staff. World Bank. Washington DC.

World Bank, 1996. Bolivia: Poverty, equity and income: Selected policies for expanding earning opportunities for the poor. Washington, DC: World Bank.

WRc, 2001. The 2001 International Benchmarking Review. WRc. UK.

Wetenschapplijke Raad voor het Regeringsbeleid, 2000. Het borgen van het publiek belang. Rapporten aan de Regering nr. 56. Den Haag. Sdu Staatsuitgerverij.

W.S. Atkins Ireland, 2000. National Water Study Ireland.

Yin, R.K., 2003. Case study research, design and methods, 3rd ed. Newbury Park: Sage Publications.

Zabel, T.F., Andrews, K. and Y. Rees, 1998. The Use of Economic Instruments for Water Management in Selected EU Member Countries Water and Environment Journal 12 (4), 2; 68–272.

Zahra, S.A. and J.A. Pearce, 1990. Research evidence on the Miles-Snow Typology. Journal of Management, Vol. 16, No. 4, 751-768

Zahra, S.A., 1987. Corporate Strategy Types, Environmental Perceptions, Managerial Philosophies, and Goals: An Empirical Study. Akron Business and Economic Review, 18, 64-77.

Zhang, Y-F., Kirkpatrick, C. and D. Parker, 2003a. Electricity Sector Reform in Developing Countries: an economic assessment of the effects of privatisation, competition and regulation. Discussion paper no. 31, Manchester: Centre on Regulation and Competition, Institute for Development Policy and Management, University of Manchester.

Zhang, Y-F., Kirkpatrick, C. and D. Parker, 2003b. Competition, regulation and privatisation of electricity generation in developing countries: does the sequencing of reforms matter? Mimeo, Manchester: Centre on Regulation and Competition, Institute for Development Policy and Management, University of Manchester.

Zocchi, R. and A. Mangano, 2002. Desk-Study Assessment of the Institutional Reform in the Italian Water Sector. Lab services Research Innovation – Gruppo Acea. Rome, June 2002

Strategy and performance of water supply and sanitation providers.
Effects of two decades of neo-liberalism

Dutch Summary
(Nederlandstalige Samenvatting)

Nederlandstalige samenvatting van het proefschrift

Dit proefschrift betreft de effecten van de neoliberale institutionele veranderingen in de water en sanitatie sector.

Beleidsmakers zien zich voor de taak gesteld om te kunnen voorzien in de voortdurend groeiende vraag naar water en sanitatie. Verschillende factoren bemoeilijken deze taak voor beleidsmakers. Beslissingen betreffende water en sanitatie dienstverlening hebben een groot publiek belang en bij het uitvoeren van de dienstverlening treden er allerlei externaliteiten op. Daarnaast is water en sanitatie als dienst onderhevig aan interpretatie. Het blijkt ambiguïde te zijn of het een private of een publieke dienst is, of het een monopolie is, en of het een economisch goed is.

In de afgelopen twee decennia is de belangrijkste institutionele verandering voor de sector het neo-liberalisme geweest. Sinds het begin van de negentiger jaren van de vorige eeuw, heeft het neo-liberalisme haar invloed laten gelden door middel van het propageren van meer concurrentie en private participatie. Neo-liberalisme manifesteert zich in de water en sanitatie sector in drie vormen: een verandering van eigendom van het waterbedrijf (privatisering), de introductie of toename van competitie (liberalisering), en samenwerkingsovereenkomsten met private partijen (private sector participatie).

Onderzoekers zijn er nog niet in geslaagd sluitend bewijs te produceren of deze neo-liberale institutionele veranderingen een toegevoegde waarde hebben. Derhalve is voor dit proefschrift gekozen voor een alternatieve onderzoeksopzet. De belangrijkste innovatie van het proefschrift is om zowel de (veranderende) instituties als het gedrag van de waterbedrijven in acht te nemen om eventuele veranderingen in prestaties te verklaren. Het gedrag van watervoorzieningsbedrijven wordt geïnterpreteerd in het proefschrift middels de strategieën van deze bedrijven. De belangrijkste onderzoeksvraag is in hoeverre neoliberale institutionele veranderingen een invloed hebben op de strategieën en de prestaties van waterbedrijven. Om de multi-dimensionale aard van strategieën te operationaliseren, is er een analytisch kader ontwikkeld dat een onderscheid maakt tussen wat een waterbedrijf kan doen (strategische context), wat het zou willen doen (strategisch plan), en wat het doet (strategische acties).

Om een inventarisatie te maken van de oorzaken en het belang van neo-liberale veranderingen is er een casestudie uitgevoerd van de West Europese water en sanitatie sector. Hierbij zijn de veranderingskrachten, de huidige institutionele opzet en plausible toekomst scenario's in kaart gebracht. De Europese water en sanitatie sector blijkt een gefragmenteerd landschap te zijn, erg afhankelijk van de lokale omstandigheden. De case studie toont aan dat de sector op een kruispunt is beland waarbij vele krachten de toekomstige richting bepalen. De analyse indiceert een duidelijke tendens van overheden om hun water en sanitatie dienstverlening te delegeren en private partijen bij de dienstverlening te betrekken. Wat dat betreft bevestigt de analyse dat neoliberalisme een grote invloed heeft gehad op de instituties van de water en sanitatie sector, en dat deze invloed waarschijnlijk zal voortduren in de nabije toekomst.

Uit de analyse blijkt dat neoliberale institutionele veranderingen een effect hebben op wat waterbedrijven *kunnen* doen (strategische context). Door middel van een case studie waarbij een vergelijking wordt gemaakt tussen de regelgeving in Engeland en Nederland relevant voor water bedrijven. Uit de case analyse blijkt dat behalve dat de regelgeving verschillend is voor Engeland en Nederland, dat ook de vrijheid van handelen verschillend is voor waterbedrijven. Managers van private waterbedrijven hebben meer vrijheid van handelen in hun markt- en produktstrategieën. Daar tegenover staat dat ze meer beperkingen hebben in het vaststellen van hun prijsstrategieën. De vrijheid van handelen wat betreft hun interne en externe strategieën blijkt echter weinig te verschillen.

De analyse in hoeverre private waterbedrijven andere strategieën *willen* voeren (strategische plannen) ten opzichte van hun publieke collega's is uitgevoerd middels een case studie in de Nederlandse Antillen en heeft een vergelijkbare conclusie als voor de strategische context. De case studie toonde aan dat eigendomsverhoudingen van waterbedrijven een invloed hebben op hun strategische plannen. Voor alle vijf strategische onderdelen uit het onderzoek blijkt dat de strategische plannen van het private waterbedrijf meer opportunistisch, aggressief en risicovol zijn.

De conclusies voor wat betreft de strategische context en de strategische plannen worden gedeeltelijk bevestigd in het deelonderzoek naar de strategische akties (wat de bedrijven echt *doen*). Strategische akties zijn onderzocht door een vergelijking te maken van de antwoorden van 96 senior managers van 49 waterbedrijven in verschillende Europese landen. Uit deze vergelijking blijkt dat voor alle vijf strategische onderdelen statistisch significante verschillen zijn tussen waterbedrijven met andere eigendomsverhoudingen.

Uit de analyses van welke strategie een waterbedrijf kan uitvoeren, wil uitvoeren en echt uitvoert, blijkt dat instituties de strategieën van waterbedrijven beïnvloeden. Om de analyse te completeren wordt getracht om de strategieën te koppelen aan de prestaties van de waterbedrijven. Hiervoor wordt een statistische analyse gemaakt van de scores uit het onderzoek naar de strategische akties en prestatie indikatoren die worden gebruikt in benchmarking projecten in Engeland en Nederland. Deze analyse levert enkele statistisch significante verbanden op tussen strategie en prestaties, maar te weinig om van een sterke relatie te kunnen spreken. Het vergelijken van de prestaties van waterbedrijven in de onderzoeksopzet blijft dus problematisch.

De moeilijkheid om prestaties van waterbedrijven te interpreteren, te meten en te vergelijken, compliceert een goed inzicht in de waarde van institutionele veranderingen. Deze conclusie bevestigt één van de uitgangspunten van het proefschrift dat het belangrijk is om meer onderzoek te verrichten naar het gedrag van waterbedrijven in plaats van de prestaties van deze bedrijven. Als men weet dat waterbedrijven zich anders gedragen vanwege (neoliberale) institutionele veranderingen, dan geeft dit het begin van een antwoord of waterbedrijven beter gaan presteren. Deze redenering opent vele nieuwe wegen voor relevant onderzoek naar de relatie tussen institutionele veranderingen en gedrag van waterbedrijven, en de relatie tussen gedrag van waterbedrijven en hun prestaties.

Strategy and performance of water supply and sanitation providers.
Effects of two decades of neo-liberalism

Annexes

Annex 1: Overview Empirical Enquiries Strategies and Performance

Research	Performance Indicators	Collection	Result
Snow and Hrebiniak (1980)	• ratio of total income to total assets	Objective	• Defenders, Analyzers and prospectors perform equally well and are superior to reactors, in three of four industries. • In the highly regulated industry the Reactor outperformed the other three archetypes.
Hambrick (1983)		Objective	• Defenders outperform prospectors in non-turbulent environments. • Prospectors outperform defenders in turbulent environments.
Smith et al. (1986)	• Sales growth • Profits • return on total assets • overall performance	Subjective	• Defenders, Analyzers and prospectors outperform reactors. • Small defenders outperform Analyzers and prospectors. • Prospectors outperform medium to large firms. • Analyzers outperform large firms.
Conant et al. (1990)	Profits	Objective	Defenders, prospectors and Analyzers perform equally well and outperform reactors.
Thomas et al. (1991)	• ROI • Market share	Objective	Better co-alignment between archetype and executive characteristics results in better performance.
Sriram and Anikeeff (1991)	Sales per employee	Objective	No significant difference in performance between Defenders, Analyzers and Prospectors.
Bahaee (1992)	• Load factor • Profit	Objective	Reactors under perform compared to Defenders, Analyzers and Prospectors.
Parnell and Wright (1993)	• Sales growth • Return on assets	Objective	• Defenders, prospectors and Analyzers outperform reactors. • Prospectors are best performing in terms of sales growth. • Analyzers are best performing in terms of return on assets.
Thomas and Ramsawamy (1994)	• sales, • return on assets • return on equity,	Objective	Organizations that align executive characteristics with the requirements of strategic archetypes perform significantly better than others that do not achieve such alignment.
Obert and Spencer (1996)	quality	Subjective	Significant relationship between strategy type and quality management approach.
Giminez (2000)	• Sales growth • Number of employees	Subjective	Defenders, Analyzers and prospectors outperform reactors especially in terms of sales growth.
Croteau and Bergeron (2001)	• Sales growth • Market share • Financial liquidity position	Subjective	• An outward technological profile contributes directly to organizational performance for the Analyzer strategic archetype, • An inward profile of technological deployment contributes indirectly to

Research	Performance Indicators	Collection	Result
			organizational performance for the Prospector archetype.
Truch and Bridger (2002)	• Overall performance in the last year, • return on investment in the last three years, • Growth in volume of sales in the last three years.	Subjective	A better alignment between knowledge orientation and strategy orientation results in better performance
Brown and Iverson (2004)	• Outcome performance • Goal attainment • Quality of services • growth	Subjective	Both Prospector and Defender organisations reported on average, higher levels of performance than Reactors and Analysers.
Andrews *et al.* (2005)	Comprehensive Performance Assessment	Objective	• Organizational performance is associated positively with a prospector stance and negatively with a reactor stance. • Local authorities that seek new markets for their services are more likely to perform as well.
Slater *et al.* (2006)	• customer satisfaction, • customer value, • customer retention, • sales growth, • market share, • profit.	Subjective	• Prospector performance benefited from a clearly articulated mission, while Analyzer performance was harmed by it. • Prospectors and Analyzers were both benefited from comprehensive alternative evaluation, and none of the strategic types were harmed by it. • Analyzers were the only strategic type whose performance was enhanced by situation analysis. • Prospector performance was harmed by a formal strategic formation process, while low-cost Defenders and differentiated Defenders' performance benefited from it.
DeSarbo *et al.* (2005)	• Profit margin • ROI • Market share • Customer retention • Sales growth • ROA	Objective	• By including strategic capabilities, environmental uncertainty , and performance results is a somewhat different classification that varies from Miles and Snow • Asian-based prospecting firms with technology strengths, and group 3: US based firms with market linking and management strengths were the lowest performing • Defensive firms with marketing skills, and Balanced prospecting firms were the best performing

Annex 2: The Miles and Snow Typologies described along the strategic components of Boyne and Walker

		Prospector	Analyzer	Defender	Reactor
'Market' variables					
1	Connections	Has a precise goal of expansion of connection inside service area	Evaluates possibility of expansion of service inside service area after cost/benefit analyses	Focuses its efforts on current customers and doesn't think of service expansion	Decides on service expansion based on legal obligation with the regulators
2	Inset appointment	Provides service to large customers outside service area	Has a customer base mainly in the service area, however sometimes it also provides services outside service area	Maintains a customer base which is strictly inside its service area	Defines strategy to deal with this when other companies start offerings in its area
3	Bulk water	Pursues bulk water sales as a means of growth	Following the successful example of other water companies, bulk water sales is part of the business	Does not pursue bulk water sales at all as it is not in its core activities	Satisfies bulk water sales demands coming from the outside.
4	Mergers and acquisitions	Has stakes in other water companies outside service area as a strategy for expansion	Has interest in stakes in other service areas, when potentially good opportunities arise	Does not have stakes in other companies outside the service area	Embarks in mergers when solicited by the external environment
'Products and Services' variables		**Prospector**	**Analyzer**	**Defender**	**Reactor**
1	Quality	Assures high quality and looks for innovative means of quality improvements	Quality of drinking water and effluents are very high, but also constrained by cost factors	Quality of drinking water and effluents are well beyond legal requirements, as this is one of the company focus areas	Quality is determined by regulation requirements
2	Product portfolio	Offers other services along with water and sanitation (like consultancy, education programs, etc)	Provides traditional services and innovation is limited to part of the current process.	Provides only traditional water and sanitation services	Offers other services to maintain accountability to regulators and to adapt to external trends
'Seeking Revenues' variables		**Prospector**	**Analyzer**	**Defender**	**Reactor**
1	Profit	Main investments areas are in innovation, R&D, new technologies introduction	Investments are balanced between O&M and new technologies	Main investment areas are in operation and maintenance	Main investments are in contingencies related to the problems of the moment
2	Efficiency	Main objectives are innovation and expansion.	Cost control is important but not the ultimate goal.	One of the main goal is cost control and reduction	No specific cost control activity in place
3	Asset management	Infrastructures are mainly managed based on current vs. available technologies (i.e. old technologies are replaced)	Asset management is used to improve efficiency, but also technology aging is a factor	Asset management is used to control costs associated to infrastructure maintenance and renewal	Asset management is not used to manage infrastructures
4	Gratuities	Water is provided for free to some customer to provide good image to the company	Exceptionally water is provided for free in few cases	The policy is not to provide water for free to any customer	There is no internal policy on gratuities for specific customers

'Internal Organisation' variables	Prospector	Analyzer	Defender	Reactor
1 Innovation	New technologies are clearly considered as the best means for improvement	New technologies are introduced once they show high potentials	Current technologies are sufficient to run the current services	New technologies and innovation are used as means to comply to regulation
2 Managers	Management team is composed by people with a wide spectrum of skills	Management team is in some departments specialized, in other more flexible	Management team is specialized	Management is not stable and depends also on external influences
3 Training	Systematically educational programs are offered	Sporadically educational programs are offered to employees	There is no need to gather additional skills through educational programs	No policy on educational undertakings is specified
4 Strategy formulation	Long term objectives are decided by top level managers in consultation with lower level managers and teams	Long term objectives are decided by top managers and also by middle level managers, depending on the context	The long term objectives are decided only by the top managers	There is no specific rule on who is responsible for long-term objective definition
5 Decentralization	Decision making is decentralized	Decision making is centralized, but some decisions are left at lower levels	Decision making Is centralized	Decisions are taken at upper or lower levels depending on circumstances
6 Autonomy	Each employee uses personal methods in task fulfillment	Rules and procedures are existing but there is a certain flexibility in task fulfillment	Rules and procedures define employee way of fulfilling their tasks	Employee autonomy in fulfilling their tasks depend on circumstances
7 Marketing	Marketing activity is very important	Limited marketing activity	No marketing activity	Marketing would be implemented in case external pressure impose this step
'External Organisation' variables	Prospector	Analyzer	Defender	Reactor
1 Environment	Water resources management improves the image of the company by showing care of environmental protection	Water resources management improves service quality and also positively affects company image	Water resources management as a guarantee of increased service quality	Management of water resources is performed according to existing regulations
2 Suppliers	Outsourcing is done when additional benefit can come from external know-how	Outsourcing is used to gain in efficiency and use external skills	Outsourcing is used mainly to minimize cost of operations	There is no specific policy on outsourcing
3 Benchmarking	Benchmarking is a recurrent activity	Sporadically benchmarking activity is performed	No benchmarking activity is performed	Benchmarking is performed when requested by management oversight
4 Regulator	Regulator poses important constraints on company activities	Regulator usually does not pose constraints unless decisions beyond current service scope are taken	Regulator does not pose constraints on normal activities	The scope of the services offered is mostly defined by the regulator
5 Partners	Undertaking partnership is one of the most common ways used to share best practices	Partnerships are sometimes arranged in cases where there is an obvious benefit	Partnership is not a practice since all skills are available internally	Partnerships are arranged whenever legal and/or political reasons call for it

Annex 3: Overview regional benchmarking initiatives

Region	Description
South America	The regulatory bodies in Brazil and Peru are using core indicators as part of their oversight activity. The South American national regulatory bodies have formed ADERASA (Asociacion de Entes Reguladores de Agua y Saneamiento de las Americas) bringing together Argentina, Bolivia, Colombia, Costa Rica, Chile, Nicaragua, Panama and Peru, with the intention to compare data.
Asia	Since 1993, the Water & Sanitation division of the Asian Development Bank has been carrying out metrics benchmarking by collecting and dissemination utility performance and management practices through publishing a 'Utilities Data Book'. The Book provides information on comparative performance of 50 utilities in 31 of the bank's member countries. Some of the areas covered include efficiency indicators, financial indicators, service coverage, water losses, supply reliability, result of consumer surveys, and extent of private sector participation.
Africa	Since 1996 the Water Utility Partnership for Capacity Building – Africa (WUP) has been comparing utility performances. The project includes 21 water utilities from 15 countries across Africa. As utility performance indicators function: water production, network length, and level of metering, manpower levels in the different sections of production, distribution and administration, operating expenses, broken down into power costs, manpower, chemical an other, gearing (debt-equity) ratios, asset values, collection efficiency/revenue collection in different consumer categories and tariffs, supply availability, number of connections, service coverage, and the extent/areas of involvement of the private sector.
World-wide	Since 1998, the World Bank has developed a Start-Up kit for benchmarking and is establishing an internet database for benchmarking data. The idea of the World Bank is that such digital platform responds to the call for international benchmarking, as it finds an increased similarity of problems and challenges internationally due to globalisation. The World Bank Start up kit identified several core indicator categories. OfWat has initiated since 1996 an international comparative assessment reflecting the performances of the English and Welsh water and service providers with water and service providers in Europe, the USA and Australia.
Scandinavia	In 1995, six Scandinavian cities of Copenhagen, Oslo, Helsinki, Stockholm, Malmö and Gothenburg, started a metric benchmarking initiative between the water providers in the respective cities. The project focused on customer survey/satisfaction, quality of service, supply availability, environmental protection, organizational/personnel matters and economy/efficiency.

Annex 4:
Standard performance indicators water service providers

No	Core indicator category	Indicator	Unit	Concept
1	Coverage	Water coverage	%	Population with easy access to water services / total population under utility's responsibility.
2	Water consumption and production	Water production	Lpcd m3/conn/m m3/hh/m	Total annual water supplied to the distribution system expressed by population served per day; per connection per month; and by household per month
		Water consumption	Lpcd M3/conn/m M3/hh/m	Total annual water sold expressed by population served per day; per connection per month; and by household per month
		Metered water consumption	Lpcd M3/conn/m M3/hh/m	Total annual metered water consumed expressed by population served per day; per connection per month; and by household per month
3	Unaccounted for water	Unaccounted for Water	%	Difference between water supplied and water sold expressed as a percentage of net water supplied
4	Metering practices	Proportions of connections that are metered	%	Total number of connection with operating meter / total number of connections
		Proportion of water sold that is metered	%	Volume of water sold that is metered / total volume of water sold.
5	Pipe network performance	Pipe breaks	Breaks/km/yr Breaks/conn/yr	Total number of pipe breaks per years expressed per km of the water distribution network; and per number of water connections
6	Cost and staffing	Unit operational cost	$/m3 sold $/m3 produced	Total annual operating expenses / total annual volume sold Total annual operating expenses / total annual water produced
		Staff per 1,000 water connections Staff per 1,000 consumers	#	Total number of staff expressed as per thousand water connections and water population served
		Labour costs as a proportion of operating costs	%	Total annual labour costs expressed as a percentage of total annual operational costs
		Contracted out service costs as a proportion of operational costs	%	Total costs of service contracted out to the private sector expressed as a percentage of total annual operational costs
7	Billing and collections	Average tariff of water	$/m3/yr $/conn/yr $/hh/yr	Total annual operating revenues expressed by amount of water sold; by number of connections; and by households served
		Total revenues per population served/GDP	%	Total annual operating revenues per population served / National GDP per capita.
		Ratio of industrial to residential charges	%	The average charge (per m3) to industrial customers compared against the average charge (per m3) to residential customers
		Connection charge	$ and % GDP	The cost to make a residential pipe connection to the water system measures in absolute amount and as a proportion of national GDP per capita
8	Financial performance	Working ratio		Total annual operating expenses / total annual operating revenues
		Debt service ratio	%	Total annual debt service expressed as a percentage of total annual operating revenues
9	Capital investment	Investment	% operating revenues $/c	Total annual investments expressed as a percentage of total annual operating revenues; and per capita served
		Net fixed assets / capita	$/capita	Total annual net fixed assets per capita served.

Annex 5: Overview of regulatory regimes UK and the Netherlands

Public policy / Water Cycle	Policy objectives		Instruments (prescriptive, incentive, informative, self-regulative)	
	England & Wales	The Netherlands	England & Wales	The Netherlands
1. Resource access	Prevent over-abstraction and improve control over the environmental effects of water abstraction. Ensure a fair and efficient allocation of water between competing local demands.	Preserve/improve quality of potential surface drinking water resources and protect them against the pollution. Manage the drinking water resource access.	*Pr.* Abstraction licences. Water Protection zones. Sites of Special Scientific Interest *Inc.* In case a licensed abstraction causes damage or loss to anyone, he is entitled to compensation from the abstractor. *Inf.* Applications for abstraction licenses need to be brought to the attention of those likely to be affected.	*Pr* Permits required for withdrawals or abstractions. Limitation of activities inside the prevention area *Inc.* Fees on groundwater withdrawals per m3
2. Production and Distribution	Ensure efficient provision of good quality drinking water and compliance with national and EU regulations at affordable prices for consumers through quality standards, and transparent public information policies	Contribute to public health improvement by providing good quality water in adequate but not excessive amounts.	*Pr..* Quality standards. Monitoring the quality of drinking water supplied. Water companies have to submit work programmes that outline how they will manage to comply with standards. Sanctions. *Inc.* Water price. *Inf.* Publishing the results of monitoring. *Self-reg.* Water companies are to self-control drinking water quality	*Pr.* Drinking water quality and production requirements. Missions of public service *Inc.* Temporary closure if required, due to non compliance of standards, Public loans from NWB and BNG. Tariff setting *Inf* Awareness building, information to be provided to VROM inspectors *Self-reg* Benchmark studies held by the VEWIN. Consumer informed of the quality of drinking water
3. Sewerage and Treatment	Ensure the provision of sewerage collection and treatment through controls and monitoring devices financed through (financial) cost covering charges.	Preserve the quality of surface and groundwater. Avoid problems of dilution in the sewers. Improve connection rate with treatment plants. Cost recovery	*Reg.* Pollution monitoring. Discharge consent *Inc.* Sewerage charges. Charge on trade effluent discharged into the sewer.	*Pr.* Environmental and sewerage plans. Capacity norms for the communal sewer and at the water board connection point. *Inc.* Subsidies to build individual sceptic tanks. Public loans BNG. Sewerage tax/charges *Self-reg.* Introduction of comparative competition by RIONED benchmark

Source: Euromarket Workpackage 4: Country report England & Wales, Country report The Netherlands

Annex 6:
Questionnaire Strategy Survey (Dutch Version)

UNESCO-IHE
Institute for Water Education

EEN INTERNATIONAAL VERGELIJKEND ONDERZOEK NAAR STRATEGISCH GEDRAG VAN DRINKWATER BEDRIJVEN

Deze vragenlijst is onderdeel van een internationaal onderzoek onder drinkwater bedrijven in Groot Brittanië, Italië en Nederland, met de doelstelling om het strategisch gedrag beter te begrijpen. U wordt vriendelijk verzocht alle vragen te beantwoorden

Bij voorbaat dank voor uw medewerking

Afdeling Management and Institutions
UNESCO-IHE, The Institute for Water Education
Westvest 7
Postbus 3015
2601 DA Delft, Nederland

WIJ VERZOEKEN U VRIENDELIJK ALLE VRAGEN TE BEANTWOORDEN.

Het totale tijdsbeslag voor het invullen van de vragenlijst is ongeveer 10 minuten.

Uw medewerking aan dit onderzoek wordt ten zeerste gewaardeerd. Mocht u een samenvatting van de onderzoeksresultaten willen ontvangen, schrijft u dan svp. uw naam en adres op de achterkant van de envelop (NIET op de vragenlijst). Wij zullen er op toe zien dat u de resultaten dan krijgt toegezonden.

Is er nog iets anders betreffende het strategisch gedrag van uw bedrijf dat u ons wilt laten weten? Gebruik dan svp. de onderstaande ruimte hiervoor. Andere op- of aanmerkingen waarvan u denkt dat deze ons kunnen helpen in het begrijpen van strategisch gedrag van water bedrijven worden tevens gewaardeerd.

_____

UNESCO-IHE
Institute for Water Education

	oneens ◄——► eens
Vragen	*(omcirkel uw antwoord)*

INTERNE ORGANISATIE

1 Ons bedrijf loopt voorop op het gebied van innovaties. — 1 2 3 4 5 Weet niet

2 De belangrijkste reden voor ons bedrijf om te innoveren is verscherpte regelgeving. — 1 2 3 4 5 Weet niet

3 De kennis en vaardigheden van onze leidinggevenden zijn divers, flexibel en gericht op veranderingsmanagement. — 1 2 3 4 5 Weet niet

4 Buitenstaanders (zoals overheidsinstanties) hebben een grote invloed op de selectie van onze leidinggevenden. — 1 2 3 4 5 Weet niet

5 Ons bedrijf stimuleert actief bijscholing van haar medewerkers. — 1 2 3 4 5 Weet niet

6 De belangrijkste reden om onze medewerkers bij te scholen is om op de hoogte te zijn van de laatste ontwikkelingen in het vakgebied. — 1 2 3 4 5 Weet niet

7 Bij het vormgeven van onze bedrijfsstrategie zijn heel veel medewerkers betrokken op een continue basis. — 1 2 3 4 5 Weet niet

8 Het hangt sterk af van de beschikbaarheid welke medewerkers betrokken worden bij het vormgeven van onze bedrijfsstrategie. — 1 2 3 4 5 Weet niet

9 Ons bedrijf is gedecentraliseerd waarbij elke afdeling een hoge mate van autonomie heeft. — 1 2 3 4 5 Weet niet

UNESCO-IHE
Institute for Water Education

oneens ◄──► eens

Vragen
(omcirkel uw antwoord)

10 In ons bedrijf is er een tendens dat de mate van autonomie van afdelingen varieert als omstandigheden veranderen.   1 2 3 4 5   Weet niet

11 In ons bedrijf, vinden medewerkers meestal wel hun eigen weg in het oplossen van problemen and het uitvoeren van hun taken.   1 2 3 4 5   Weet niet

12 Als een bepaalde taak geïdentificeerd wordt als belangrijk, dan neemt de directie de uitvoering ervan snel over.   1 2 3 4 5   Weet niet

13 Ons bedrijf spendeert weinig geld aan marketing.   1 2 3 4 5   Weet niet

14 De belangrijkste reden dat wij geld aan marketing spenderen is om in te spelen op specifieke behoeften van toezichthouders, gemeentes of klanten.   1 2 3 4 5   Weet niet

PRODUKTEN EN DIENSTEN

15 Ons bedrijf heeft kwaliteitsstandaarden geformuleerd die de wettelijke normen ver overstijgen.   1 2 3 4 5   Weet niet

16 Verbeteringen in onze dienstverlening komen met name tot stand door interventies van onze toezichthouders   1 2 3 4 5   Weet niet

17 Naast water gerelateerde produkten en diensten, leveren wij ook vele andere produkten en diensten.   1 2 3 4 5   Weet niet

18 Alle producten en diensten die wij leveren, zijn typisch voor een normaal Nederlands waterbedrijf.   1 2 3 4 5   Weet niet

	oneens ◄────► eens
Vragen	*(omcircel uw antwoord)*

KLANTEN

19 Wij vinden het belangrijk om zoveel mogelijk nieuwe aansluitingen te realiseren, zodat we groter kunnen worden.   1 2 3 4 5   Weet niet

20 Wij maken nieuwe aansluitingen omdat we daartoe wettelijk verplicht zijn.   1 2 3 4 5   Weet niet

21 Ons bedrijf is actief bezig om klanten buiten ons voorzieningsgebied te benaderen om onze diensten aan te verlenen.   1 2 3 4 5   Weet niet

22 Als wij merken dat een ander waterbedrijf een van onze klanten benaderd heeft, dan reageren wij snel.   1 2 3 4 5   Weet niet

23 Ons bedrijf heeft een actieve strategie om meer bulk water aan andere waterbedrijven te leveren.   1 2 3 4 5   Weet niet

24 Ons bedrijf levert bulk water aan naburige waterbedrijven alleen als we daar nadrukkelijk tot verzocht worden.   1 2 3 4 5   Weet niet

25 Ons bedrijf is actief op zoek om te fuseren met andere waterbedrijven.   1 2 3 4 5   Weet niet

26 Indien ons bedrijf zou fuseren met een ander waterbedrijf is dat alleen omdat we niet anders kunnen.   1 2 3 4 5   Weet niet

EFFICIENCY

27 Het overgrote deel van onze uitgaven gaat naar het beheer van al bestaande infrastructuur.   1 2 3 4 5   Weet niet

28 In het geval dat wij winst maken, dan wordt deze winst gebruikt voor datgene wat op dat moment de hoogste prioriteit heeft.   1 2 3 4 5   Weet niet

UNESCO-IHE
Institute for Water Education

	oneens ◄────► eens
Vragen	*(omcirkel uw antwoord)*

29 Wij streven zeer nadrukkelijk ernaar onze efficiency te vergroten.   1 2 3 4 5 Weet niet

30 Efficiency verbeteringen die wij realiseren, zijn voor een belangrijk deel toe te schrijven aan interventies van buitenstaanders (zoals toezichthouders, gemeentes, provincies).   1 2 3 4 5 Weet niet

31 Wij maken gebruik van uitgebreide 'Asset Management' plannen, zodat we continue de toestand van onze infrastructuur kunnen controleren.   1 2 3 4 5 Weet niet

32 Ons bedrijf heeft geen geformaliseerde en gestructureerde wijze van 'Asset Management'.   1 2 3 4 5 Weet niet

33 In geval dat wij gratis water leveren, doen wij dat omdat dat ten goede komt aan ons imago.   1 2 3 4 5 Weet niet

34 Als wij gratis water leveren, is dat omdat wij daartoe wettelijk gedwongen zijn.   1 2 3 4 5 Weet niet

EXTERNE RELATIES

35 Wij geven veel aandacht aan milieu beheer omdat dat ons imago ten goede komt.   1 2 3 4 5 Weet niet

36 Wij doen veel meer aan milieu beheer dan datgene wat van ons wettelijk vereist wordt.   1 2 3 4 5 Weet niet

37 Wij trachten onze afhankelijkheid van toeleveranciers te minimaliseren door zelf alle benodigde kennis en kwalificaties in huis te hebben.   1 2 3 4 5 Weet niet

38 Momenteel hebben we geen duidelijk overzicht van al onze toeleveranciers noch een strategie hoe met hen om te gaan.   1 2 3 4 5 Weet niet

oneens ◄────► eens

Vragen   *(omcirkel uw antwoord)*

39 Wij verwelkomen 'benchmarken' met andere waterbedrijven omdat ons dat de mogelijkheid geeft om van anderen te leren.   1 2 3 4 5 Weet niet

40 De belangrijkste reden voor ons om deel te nemen aan de VEWIN benchmark is om niet achter te blijven bij de andere waterbedrijven.   1 2 3 4 5 Weet niet

41 Ons bedrijf ziet toezichthouders (zoals ministeries en gemeenten) als beperkend voor hoe wij eigenlijk ons bedrijf zouden willen uitvoeren.   1 2 3 4 5 Weet niet

42 Toezichthouders (zoals ministeries) bepalen in hoge mate hoe wij ons bedrijf uitvoeren.   1 2 3 4 5 Weet niet

43 Ons bedrijf vind het uitermate belangrijk om samenwerkingsverbanden op te zetten met diverse partijen.   1 2 3 4 5 Weet niet

44 Meestal gaan wij alleen samenwerkingsovereenkomsten met anderen aan als wetgeving dat noodzakelijk maakt.   1 2 3 4 5 Weet niet

45. Voor welk water bedrijf werkt u?

Naam bedrijf: ....

46. Wat is uw positie in het bedrijf?

☐   Directie
☐   Midden management
☐   Anders

47 **In welke van de onderstaande beschrijvingen herkent u de meeste overeenkomsten met uw eigen drinkwater bedrijf?**

Kruis svp. de juiste box aan.

Type A: Dit water bedrijf streeft ernaar continue haar markten uit te breiden dmv. het benaderen van nieuwe klanten en het aanbieden van nieuwe diensten. Innovatie en de toepassing van de allernieuwste technieken worden hogelijk gewaardeerd. De interne organisatie is flexibel, gedecentraliseerd en creativiteit wordt gestimuleerd. Ook extern is het bedrijf erg actief, voortdurend op zoek naar nieuwe ideeën, partners en mogelijkheden.

Type B: Dit water bedrijf is volledig toegewijd aan het leveren van water aan haar huidige klanten. Dit doet het goed en het is trots op de uitstekende kwaliteit van haar dienstverlening. Alle inkomsten worden geïnvesteerd in het verder verbeteren van het productie en distributie proces. Het bedrijf concentreert zich om haar klanten in haar bedieningsgebied zo goed mogelijk te bedienen. De rest van wereld wordt alleen belangrijk als die implicaties geeft voor deze taak.

Type C: Dit water bedrijf onderneemt wel nieuwe activiteiten maar niet alvorens goed alle voor en nadelen te analyseren. Ideeën worden eerst onderworpen aan een gestructureerde risico analyse, waarbij de ervaringen van andere bedrijven worden gewogen. Deze strategie heeft het bedrijf in het verleden een gestadige en geleidelijke groei gegarandeerd.

Type D: Dit is een water bedrijf dat prioriteit legt in het snel anticiperen op externe ontwikkelingen. Wanneer een regulerende instantie veranderingen wil gaan aanbrengen dan springt het bedrijf hier snel op in. Dit is noodzakelijk want de invloed van politiek en overheid op het bedrijf is groot.

Annex 7: Comparing self-typing and multi-item measurement

Operators	Prospector (P) – Analyser (A) – Defender (D) continuum					Non Reactor – Reactor Continuum					Aggregated PAD	Aggregated NR-R	Self-description
	Market	Product/Services	Price/Efficiency	Internal organisation	External organisation	Market	Product/Services	Price/Efficiency	Internal organisation	External organisation			
Albion Water Ltd.	63	38	75	71	44	-	25	25	32	31	60:P	25	PPP
Anglian Water Company	50	25	19	43	50	44	88	25	46	30	40:D	42	D
Bristol Water Ltd	25	50	6	64	55	44	75	63	38	45	43:D	47	D
Cambridge Water Company Plc	58	50	25	50	65	50	63	25	21	35	50:A	35	A
Seven Trent Water Company	56	38	14	57	55	44	54	44	45	22	47:A	40	PAD
Southwest Water	50	-	31	68	80	38	63	58	68	55	58:P	57:R	D
Sutton & East Surrey Water Plc	34	44	19	63	43	59	69	66	63	25	43:D	55:R	PA
United Utilities	56	44	31	63	53	56	63	59	59	43	52:A	55:R	AA
Wessex Water	38	50	38	57	50	63	75	50	68	50	48:A	61:R	D
Northern Ireland Water company	6	38	44	36	80	69	88	81	50	35	42:D	59:R	DDD
Scottish Water	19	25	42	50	56	67	88	33	67	55	41:D	61:R	D
Vitens	44	13	27	63	67	55	50	13	38	15	49:A	32	PD
Hydron Flevoland	50	38	25	68	60	50	25	6	50	20	52:A	33	D
Waterleiding bedrijf Amsterdam	48	17	49	55	63	42	50	47	43	38	51:A	43	AAD
Brabant Water	58	46	35	43	57	57	21	40	36	30	48:A	38	AAD
Evides	69	50	47	63	43	56	25	41	46	35	55:P	43	AA
WML	63	31	47	73	70	44	50	19	29	45	62:P	35	AD
DZH	67	50	19	57	70	56	25	50	54	40	54:P	48	D
PWN	8	21	25	52	67	71	25	33	44	27	40:D	41	DDD
WMD	50	6	47	55	65	47	38	38	52	38	51:A	44	PP
Hydron Midden-Nederland	60	25	40	56	75	56	50	40	54	20	55:P	44	AAA
AQP S.p.A.	34	50	28	36	48	53	75	53	55	40	38:D	53:R	D
CAM S.p.A.	69	13	13	75	55	25	50	31	75	55	52:A	51:R	P

Operators	Prospector (P) – Analyser (A) – Defender (D) continuum					Non Reactor – Reactor Continuum					Aggregated PAD	Aggregated NR-R	Self-description
	Market	Product/ Services	Price/ Efficiency	Internal organisation	External organisation	Market	Product/ Services	Price/ Efficiency	Internal organisation	External organisation			
Valle Umbra Servizi S.p.A.	75	50	25	60	69	50	50	44	75	55	54:P	58:R	D
ACA S.p.A.	42	50	33	36	60	46	63	58	69	48	43:D	58:R	AD
Gran Sasso Acque S.p.A.	50	38	3	73	50	38	81	41	80	38	48:A	56:R	PD
Tennacola S.p.A.	63	38	19	71	35	44	50	44	61	40	49:A	49	D
S.A.S.I. S.p.A.	38	38	25	36	50	31	75	63	50	40	38:D	49	D
Acque del Chiampo S.p.A.	38	38	25	50	45	50	50	50	54	45	41:D	50:R	D
SMAT S.p.A.	42	42	19	75	68	54	75	25	46	32	54:P	43	PPP
AMAG S.p.A.	63	38	6	61	75	63	63	31	64	40	52:A	52:R	A
Servizio Idrico Integrato Terni	38	38	27	52	47	45	63	48	61	35	42:D	50:R	AR
Acqualatina S.p.A.	47	38	16	66	53	41	69	50	64	40	48:A	52:R	PR
Metropolitana Milanese S.p.A.	44	38	6	68	55	50	63	63	67	40	47:A	56:R	D
Acquedotto Lucano S.p.A.	38	44	25	61	45	58	69	56	59	43	45:A	55:R	D
Salerno Sistemi S.p.A.	50	25	25	61	65	75	63	75	61	40	50:A	61:R	D
Umbra Acque S.p.A.	47	38	16	63	58	34	63	34	66	40	48:A	48	PA
Acquedotto del Fiora S.p.A.	23	40	33	51	61	65	75	51	64	33	44:D	56:R	AAADD
Russo Servizi S.p.A.	69	63	31	61	65	56	75	63	61	55	58:P	60:R	D
Acque S.p.A.	38	25	-	83	65	31	100	38	75	65	49:A	58:R	P
CIIP S.p.A.	69	38	28	46	58	47	44	53	70	28	49:A	51:R	AA
Publiacqua S.p.A.	38	38	6	75	55	44	50	56	54	30	48:A	47	P
ASM Brescia S.p.A.	78	56	13	68	68	31	31	28	39	20	59:P	31	PA
Brianza Acque S.p.A.	58	-	25	60	25	50	-	50	60	25	50:A	53:R	D
BIM Gestione Servizi Pubblici S.p.A.	38	56	50	46	63	44	63	56	57	53	50:A	54:R	PA
HERA S.p.A.	75	63	19	57	70	31	75	50	71	45	57:P	55:R	P
Gorgovivo Multiservizi S.p.A	81	75	31	57	70	67	75	50	75	45	61:P	61:R	A
ASP S.p.A.	47	75	22	70	53	46	44	53	50	53	53:A	50:R	AA

Annex 8: Overview Outcome ANOVA analysis

Table 1 Relation between ownership types and the market strategic actions

Variable	scales	Ownership			Uni-variate F-value	P (probability)
		Private n=10 M(SD)	**Mixed n=16 M(SD)**	**Public n=23 M(SD)**		
1. Connections	Prospector	67 (17)	73 (19)	63 (24)	1.04	0.36
	Reactor	**44 (24)**	**38 (18)**	**60 (33)**	**3.17**	**0.05***
2. Inset appointments	Prospector	48 (31)	29 (24)	31 (25)	1.78	0.18
	Reactor	**75 (10)**	**42 (28)**	**55 (30)**	**3.79**	**0.03***
3. Bulk water	Prospector	41 (20)	30 (17)	47 (31)	1.99	0.15
	Reactor	50 (24)	58 (27)	56 (27)	0.26	0.77
4. Mergers and acquisitions	**Prospector**	**27 (24)**	**53 (27)**	**61 (35)**	**4.10**	**0.02***
	Reactor	19 (18)	45 (26)	37 (34)	2.24	0.12
Combined	*Prospector*	*49 (27)*	*47 (28)*	*50 (32)*	*0.387*	*0.68*
	Reactor	*47 (27)*	*46 (25)*	*52 (32)*	*0.996*	*0.37*

Table 2 Relation between ownership type and the strategic products/services actions

Variable	Scales	Ownership			Uni-variate F-value	P-value (probability)
		Private n=10 M(SD)	*Mixed n=15 M(SD)*	*Public n=23 M(SD)*		
1. Quality	Prospector	41 (25)	41 (17)	34 (25)	0.62	0.54
	Reactor	**62 (21)**	**45 (21)**	**37 (29)**	**3.32**	**0.04***
2. Product portfolio	Prospector	43 (19)	42 (24)	43 (31)	0.00	0.99
	Reactor	63 (22)	80 (14)	69 (22)	2.69	0.08
Consolidated	*Prospector*	*42 (22)*	*41 (21)*	*38 (29)*	*0.26*	*0.78*
	Reactor	*62 (21)*	*63 (25)*	*53 (30)*	*1.68*	*0.19*

Table 3 Relation between ownership and the strategic seeking revenue actions

Variable	Scales	Ownership			Uni-variate F-value	P-value (probability)
		Private n=10 M(SD)	*Mixed n=15 M(SD)*	*Public n=23 M(SD)*		
1. Profit allocation	**Prospector**	**47 (27)**	**15 (13)**	**32 (28)**	**5.70**	**0.01***
	Reactor	**58 (23)**	**76 (11)**	**46 (24)**	**9.58**	**0.00***
2. Efficiency	Prospector	10 (23)	16 (13)	15 (16)	0.45	0.64
	Reactor	**67 (32)**	**63 (13)**	**44 (34)**	**3.17**	**0.05***
3. Asset management	Prospector	23 (14)	20 (15)	34 (23)	2.70	0.08
	Reactor	19 (11)	30 (16)	38 (25)	2.73	0.08
4. Gratuities	**Prospector**	**14 (13)**	**30 (25)**	**40 (27)**	**3.81**	**0.03***
	Reactor	48 (23)	34 (28)	36 (26)	0.63	0.54
Consolidated	*Prospector*	*23 (24)*	*20 (17)*	*30 (25)*	*3.97*	*0.02***
	Reactor	*48 (30)*	*51 (26)*	*41 (27)*	*2.50*	*0.09*

Table 4 Relation between ownership and the strategic internal organisation actions

Variable	Scales	Ownership			Uni-variate F-value	P-value (probability)
		Private n=10 M(SD)	Mixed n=15 M(SD)	Public n=23 M(SD)		
1. Innovation	Prospector	55 (28)	50 (28)	59 (28)	0.49	0.62
	Reactor	46 (26)	56 (16)	48 (26)	0.67	0.52
2. Knowledge and skills	Prospector	75 (23)	72 ((17)	59 (22)	2.89	0.07
	Reactor	**25 (35)**	**66 (20)**	**37 (33)**	**6.41**	**0.00***
3. Training	Prospector	74 (12)	79 (15)	74 (19)	0.48	0.62
	Reactor	65 (25)	76 (14)	62 (21)	2.52	0.09
4. Strategy-formulation	Prospector	65 (22)	52 (24)	49 (22)	1.97	0.15
	Reactor	**41 (26)**	**66 (17)**	**44 (22)**	**5.84**	**0.00***
5. Decentralisation	Prospector	58 (21)	57 (21)	54 (18)	0.19	0.83
	Reactor	62 (19)	57 (31)	54 (21)	0.34	0.72
6. Empowerment	**Prospector**	**67 (14)**	**55 (18)**	**68 (14)**	**3.25**	**0.04***
	Reactor	63 (19)	78 (17)	68 (20)	1.30	0.28
7. Marketing	Prospector	29 (19)	52 (21)	41(29)	2.57	0.09
	Reactor	48 (22)	44 (19)	57 (26)	1.57	0.22
Consolidated	*Prospector*	*60 (24)*	*60 (23)*	*58 (24)*	*0.487*	*0.62*
	Reactor	***50 (28)***	***63 (22)***	***53 (27)***	***7.077***	***0.00****

Table 5 Relation between ownership and the strategic external organisation actions

Variable	Scales	Ownership			Uni-variate F-value	P-value (probability)
		Private n=10 M(SD)	Mixed n=16 M(SD)	Public n=23 M(SD)		
1. Environment	**Prospector**	**48 (22)**	**71 (13)**	**70 (24)**	**4.307**	**0.02***
	Reactor	39 (27)	26 (17)	28 (19)	1.448	0.25
2. Suppliers	Prospector	48 (36)	33 (17)	48 (28)	1.816	0.17
	Reactor	**22 (18)**	**51 (21)**	**40 (26)**	**4.716**	**0.01***
3. Benchmarking	**Prospector**	**79 (20)**	**54 (13)**	**75 (21)**	**7.175**	**0.00***
	Reactor	26 (22)	34 (18)	45 (30)	2.265	0.12
4. Regulator	Prospector	51 (26)	54 (19)	44 (22)	1.005	0.37
	Reactor	44 (32)	46 (19)	43 (25)	0.050	0.95
5. Partnerships	Prospector	61 (18)	67 (19)	74 (22)	1.382	0.26
	Reactor	**56 (27)**	**54 (22)**	**32 (26)**	**4.762**	**0.01***
Total	*Prospector*	*57 (27)*	*56 (21)*	*62 (27)*	*1.87*	*0.16*
	Reactor	*37 (27)*	*42 (22)*	*38 (26)*	*0.73*	*0.49*

Annex 9: Weighing factors OPA

Key area	Measure	Weighting for Water and Sewerage companies	Weighting for all water companies
Water supply	DG2	0.75	0.75
	DG3	0.75	0.75
	DG4	0.5	0.5
	Drinking water quality	1	1
	Subtotal water supply	**3**	**3**
Sewerage service	Sewer flooding incidents (capacity)	0.5	Not relevant
	Sewer flooding incidents (other causes)	0.75	
	Properties at risk of sewer flooding more than once in 10 years	0.25	
	Subtotal sewerage service	**1.5**	
Customer service	Company contact	0.75	0.75
	Other customer service	0.75	0.75
	Subtotal customer service	**1.5**	**1.5**
Environmental performance	Category 1 and 2 pollution incidents per million equivalent resident population (sewage)	0.5	Not relevant
	Category 3 pollution incidents per million equivalent resident population (sewage)	0.25	
	Sludge disposal	0.25	
	Percentage equivalent population served by STWs in breach of their consent	1	
	Category 1 and 2 pollution incidents (water)	0.25	0.25
	Leakage	0.5	1
	Subtotal environmental performance	**2.75**	**1.25**
Weighting totals		**8.75**	**5.75**

Source OFWAT, 2007

Annex 10: Curriculum Vitae Marco Schouten

Name	Schouten, Marco
Year of birth	1970
Nationality	Dutch
Organisation	UNESCO-IHE Delft, The Institute for Water Education
Present position	Deputy Head of Department Management and Institutions, Senior Lecturer Water Services Management
Years with UNESCO-IHE	Since 1 May 2003

EDUCATION

1994	MA. Business Economics, Vrije Universiteit Amsterdam, The Netherlands

EMPLOYMENT RECORD

2003 - to date:	Senior Lecturer Water Services Management; and Deputy Head of Department Management and Institutions, UNESCO IHE-Delft, The Netherlands
1999 –2003:	Business Economist, IWACO B.V. – later merged into Royal Haskoning B.V.
1998 –1999:	Management Consultant, Molenaar & Lok Consultancy, The Netherlands
1996 –1998:	Management Consultant, Ernst & Young Management Consultants, Suriname
1995 –1996:	Consultant, Management Consultancy Agency Atelier 28, Amsterdam, The Netherlands
1992 –1996:	Management Consultant, Free Lancer

KEY QUALIFICATIONS

Mr. Schouten is working as Deputy Head of the Department 'Management and Institutions' and senior lecturer on topics related to Water Service Management for UNESCO IHE-Delft. His experience encompasses institutional strengthening, strategy development, business process re-engineering, training development and delivery, finance and administration, and private sector participation in the water sector. Before he joined UNESCO IHE-Delft, Mr. Schouten gained ten yours of international consultancy experience, of which he lived and worked for five years in developing countries.

MAIN DISCIPLINE / SPECIALISATION

Economical, financial and institutional issues related to the water sector

EXPERIENCE RECORD

EXPERIENCE in CONSULTANCY, PROJECT ASSIGNMENTS and EDUCATION

2008 – date Uganda	Integrated approaches and strategies to address the Sanitation Crisis in Unsewered Slum Areas in African mega-cities Researcher
2007 – date Iran	Training and Capacity Building for Water and Wastewater Sector Acquisition and International Trainer
2006 - date The Netherlands	Faculty of Economics and Business of the Vrije Universiteit (VU) Amsterdam Guest Lecturer International Business Administration
2004 – date Southern Africa	WATERNET Management Consultant and Senior Lecturer
2003 – date The Netherlands	UNESCO IHE-Delft, educational programme Course co-ordinator / Senior Lecturer Water Services Management
2007 South Africa	Training Program in Public-Private Partnerships, Nelspruit Project Leader and acquisition
2007 Zambia	Training Program in Water Services: Reaching the Poor in Peri-Urban Areas Project Leader and acquisition
2005 – 2006 Bosnia Herzegovina, Croatia, Serbia-Montenegro, Slovenia	Pilot River Basin Plan for the Sava River Economist
2006 Namibia	Training program in Public-Private Partnerships in the Southern African Water Sector Project Leader and acquisition
2005 Indonesia	Training program in Private Provision of Infrastructure Technical Assistance Trainer
2005 Netherlands Antilles	Evaluation of proposals for an integrated sewage and drinking water system Deputy Team Leader
2004 Belgium	Feasibility Study Finance Access to Water and Water Purification Business Economist
2003 – 2006 European Union	EUROMARKET Project Project Co-ordinator and researcher
2003 Austria	Training program Water and Sanitation in developing countries Trainer
2003 Nigeria	Capacity building and Institutional Analysis for Reform Economist
2003 Hungary	Investigation on Municipal Solid Waste Landfills Economist
2002	Management Support Programme and Private Sector Participation Options

Egypt	Study of the Kafr El Sheikh Water and Sewerage Company Economist
2002 The Netherlands	Municipality of Kloosterzande aan Zee Business Economist
2002 Bosnia-Herzegovina	Water Institutional Strengthening Project Economist
2000 Southern Africa	Evaluation IT Billing and Collection System Member of Evaluation Committee
1999 –2001 Egypt	Fayoum Drinking Water and Sanitation Project Expert on Finance and Management
1998 –1999 The Netherlands	IT implementation ING Bank headquarters Implementation Co-ordinator
1996 –1998 Surinam	Quality Improvement projects for a shrimp exporting company, a flour mill and a cattle fodder plant Advisor / Project leader Administrative Organisation Improvement Projects for a bank, an oil exploration and refinery company and a department of the Ministry of Finance Advisor / Project leader Performance Improvement projects for an insurance company, a brewery, a juicy factory and the Primary Health Services Suriname Advisor / Project leader
1995 –1996 The Netherlands	Interim Management SBK Assistant Interim Manager
1995 –1996 The Netherlands	Refugees employment project Advisor
1992 –1996 The Netherlands	Various Projects for a computer trading company, the Vrije Universiteit, commercial employment agencies, a health centre, and a retail company Advisor

PUBLICATIONS

JOURNAL PAPERS

- *100-days Rapid Change Initiatives in African Public Water Utilities"*. Submitted December 2008 to International Journal of Public Sector Management. Emerald. Together with T. Buyi.
- *How Institutional Arrangements Foster Transparency, Participation and Accountability: Analysis of Scotland and Barcelona Regulatory Frameworks for Water Services Provision"*. Submitted September 2008 to Water Policy. Elsevier. Together with M. Pascual Sanz and M. Handtke Domas.
- *"Regulation and comparative discretion of publicly and privately owned water companies in the Netherlands, England and Wales"*. Accepted May 2008 in Water Policy. IWA Publishing. Together with M.P. van Dijk.
- *"Private Sector Involvement according to European Water Liberalisation Scenarios"*, 2008. International Journal of Water. Vol. 4.

No. 3/4: 180-196. Inderscience Enterprises Ltd. Together with M.P. van Dijk.

- *"A Caribbean Evaluation of Public versus Private Drinking Water Provision: the case of St. Maarten, Netherlands Antilles"*, 2008. International Journal for Water. Vol. 4, No. 3/4:258-274. Inderscience Enterprises Ltd. Together with D. Brdjanovic and M.P. van Dijk.
- *"Water as a Political Good: Implications for Investments"*, 2006. International Environmental Agreements: Politics, Law and Economics (6): 407-421. Springer. Together with K. Schwartz.
- *"Water as a Political Good: Revisiting the Relationship between Politics and Service Provision"*, 2006. Water Policy (9): 118-129. IWA Publishing. Together with K. Schwartz
- *"Strategie dei gestori del servizio idrico in Italia"*, 2006. Management delle Utilities. Maggioli Editore. Anno 4-Numero 2. Riversi Trmestrale, Umbria. Together with G. Casale.
- *"The European Waterscape of Management Structures and Liberalizatio"*, 2004. Water and Wastewater International Magazine, June 2004. Together with M.P. van Dijk
- *"Groei en ontwikkeling van organisaties. Een oeroud patroon: tegen eigen grenzen aanlopen en zichzelf overstijgen"*, 1997. M&O. Tijdschrift voor organisatiekunde en sociaal beleid, 51(4), 62-77. Together with P.J. Tack.

CONFERENCE PAPERS

- *"Exploring Strategies of Water Providers"*, 2007. Paper presentation at the 1st IWA Utilities Conference 'Customer Connection', June 2007, Maastricht, The Netherlands.
- *"Regulatory impositions posed upon strategic actions of publicly and privately owned water companies, in respectively the Netherlands and England and Wales"*, 2005. Paper presentation on the International Symposium on "Competition and stakes in the regulation of services of general interest. Feedback of the last twenty years", September 2005, Paris, France., together with Prof. M.P. van Dijk
- *"The Dynamics of the European Water and Wastewater Sector"*, 2004. Paper presentation on the International Specialty Conference on "Good Water Governance for People and Nature", September 2004, Dundee Scotland. Together with M.P. van Dijk

EDITED BOOKS

- *"Innovative Practices of African Water Providers"*, 2009. Edited by M. Schouten, E. Hes and Z. Hoko. Published by Sun Media Publishing. South Africa.

BOOK CHAPTERS

- *"The European Water Supply and Sanitation Markets"*, 2006. In Book: 'Water and Liberalisation', edited by Matthias Finger, Jeremy Allouche and Patrica Luis-Manso, Chapter 1. IWA Publishing. European Commission. Together with M.P. van Dijk, 2006

TECHNICAL REPORTS

- *"Plugging the Leak. Can Europeans find new sources of funding to fill the MDG water and sanitation gap?"*, 2005. IRC. June 2005. 54 pages. Together with C. Fonseca and R. Franceys

OTHER

- *"Over de Smaak van Nijlwater"*, 2004. Controller Magazine, April 2004:
- De Ware Tijd (national newspaper Suriname), 1997-1998: Series of monthly articles about *management*

T - #0078 - 071024 - C68 - 254/178/18 - PB - 9780415551298 - Gloss Lamination